THEORETICAL PHYSICS

Unthinkable Works !

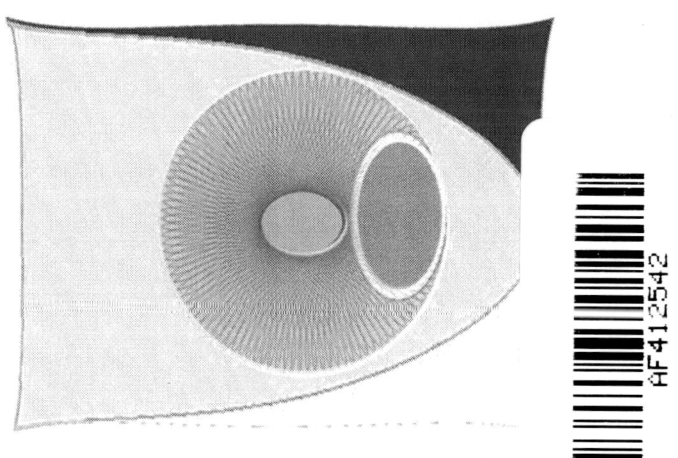

By

DEREK J. DANIEL

CHARLESTON

2010

Made and Reprinted in USA

THEORETICAL PHYSICS:
UNTHINKABLE WORKS !

A CIP catalogue for this book is available from the National Library of Ireland.

First printed and bound edition in Ireland by
Choice Publishing & Book Services Ltd, Drogheda......*February, 2009.*

Second printed and bound edition in United States of America by
CreateSpace, Charleston....................*June, 2010.*

ISBN 978 1 451584 81 3

Cover design, page editing and typesetting by Derek J. Daniel
Artwork & Illustrations: Derek J. Daniel

PREFACE

It was shortly after Christmas of the year 2007 that I realized that I was still without academic appointment, yet I had the potential to revolutionize various fields of study in theoretical physics. By late February of the year 2008 I had managed – with my somewhat meager savings – to affiliate myself with the Institute of Mathematical Sciences at the University of Malaya, where, in a space of four months, I had begun to formulate and piece together my earlier drafts, which I had originally sketched out in my own home in Australia, into three major essay works in theoretical physics, each of which contain exact solutions to problems that most academics in their own right would regard as far too intimidating to attempt. This is because the problems themselves are mainly viewed at the outset, usually by experts in their respective fields, as too impossible to solve analytically. Universally, however, if one were to deliver exact solutions to such problems and then collect them into a monograph, one would have the basis for a collection of *unthinkable works!* This monograph, therefore, is the start of my own personal collection of such unthinkable works.

In exposing to the reader the details associated with my full solution sets, I have sought to compress all major steps to reach the final result in the shortest possible path as possible, since this is my laurel toward "exactness" that I had been brought up on when elucidating or unraveling on diverse phenomena in theoretical physics. Bear in mind, though, that the exactitude involved in the mathematics covered in this monograph is in no way intended to outlast the patience of the reader.

I must add, however, that when I first approached certain reputable scientific journals for an appraisal of the essays embodied by this monograph, their main response was less than encouraging and only saw to such essays being rejected; this being without any real apparent reason, since the results contained in each essay of this monograph were neither deemed incorrect or invalid, nor disproven. Nevertheless, the essays still stand very much in accordance with their original preprints that I had started on, so the reader can safely assume that no large alterations were made, apart from a matter of re-wording one or two rather impatient sentences that appeared or sounded awkwardly perplexed in certain paragraphs.

Chapter 1 begins with a revisit of the simple harmonic oscillator as first described and solved by Erwin Schrödinger. Here the reader learns of the one factor that is totally avoided in almost all text books in quantum mechanics, the implications of which will rightfully leave the reader wondering – if not amazed – as whether or not to disqualify Erwin Schrödinger's wavefunction solution for the simple harmonic oscillator.

In Chapter 2, I have attempted to unify physics, biology and ecology into a single procession, by focusing on the Kolmogorov Master equations for birth-and-death. The truth here is that no one actually knew how to solve this kind of stochastic equation, or how to pry out a unique solution from such an equation. Accordingly, Chapter 2 is charmed with several unexpected results for the reader to observe.

Chapter 3 is my most penultimate challenge yet, since it delves right into one of the longest unsolved problems in the field of theoretical physics – well before clocks were invented – namely, the simple pendulum; the very same one that was originally observed by Galileo Galilei around 1590. The reader will find this chapter a real "cliff-hanger" as far as how straightforward the solution for the simple pendulum can be obtained without using elliptic functions.

Finally, to avoid possible conflicts of interest, the present author states that no part of the research in this monograph was in receipt of any grant support from any department or institute at the University of Malaya, nor was the author financed by his former home institution (formerly The Australian National University), which includes the Australian Government. Also, no part of this work was work made for hire. The most memorable acknowledgement, however, is dedicated to a most inspirational companion, for anonymity, named here as "Zinga", for providing me with a tremendous amount of fortitude and courage needed to pursue this type of work. This fits perfectly with the symbol that I had specifically created to meet life' toughest challenges head-on, represented by the rhinocerous and its horns, displayed on both the front cover and title page of this monograph. This symbol is also a reminder that the rhinocerous, amongst many of the other wildlife species inhabiting this planet, is truly an endangered species, so that *proceeds* from this monograph are intended as a means of raising funds for such a cause.

*October, 2008**Derek J. Daniel.*

CONTENTS

CHAPTER 1

The Quest of Finding the Missing Solutions for the Quantum Mechanical Oscillator

The issue of the angular frequency ω in Erwin Schrödinger's exact wave function for the simple harmonic oscillator approaching a zero limit, i.e., $\omega \to 0$, is a matter that is totally avoided in all textbooks in Quantum Mechanics. This is because the "true" solution for the simple harmonic oscillator should reduce to that of a *plane-wave solution*, in the limit $\omega \to 0$, a fact which the present author sets off to prove in this chapter. Here, the reader is left with the implications as to whether all calculations in quantum mechanics to date – particularly those that use simple harmonic wave functions without invoking the correct limit behaviour – are indeed *incorrect*.

1. Introduction

Schrödinger's time-independent equation for the quantum mechanical oscillator, originally called the *linear harmonic oscillator*, is defined by

$$-\frac{\hbar^2}{2m}\frac{d^2\psi}{dx^2} + \tfrac{1}{2}m\omega^2 x^2\psi = E\psi, \tag{1}$$

which, apart from a slight change in notation, coincides with the definition used in **Chapter 4** by Schiff (1965) [1], where here, ψ is the wave function, ω is the angular frequency of the oscillator, E is its energy, and h is Planck's constant which is absorbed into $\hbar = h/2\pi$.

The wavefunction solution ψ to Eq. (1) is celebrated in pp. 62 of Schiff (1965) [1] by an orthogonal class of polynomials, called the *Hermite polynomials* of degree n, denoted as (without normalization)

$$\psi = e^{-\frac{m\omega}{2\hbar}x^2} H_n\left(\sqrt{m\omega/\hbar}\,x\right). \tag{2}$$

1

However, the surprising instance with this set of polynomials is when the angular frequency ω of the linear harmonic oscillator approaches zero, $\omega \to 0$, since in this limit, Eq. (2) gives

$$\lim_{\omega \to 0} \psi = \text{constant},$$

a result that is in direct contradiction to the "true" solution. This can be verified directly from the Schrödinger wave equation, by taking the limit, $\omega \to 0$, in Eq. (1). Consequently, the solution in this limit, apart from normalization, can only be given by

$$\psi \sim e^{ikx};$$

that is, by a plane-wave with wave-number $k = \sqrt{2mE/\hbar^2}$.

In the remainder of this chapter, the original derivation presented in **Chapter** 4 of Schiff (1965) [1] will be tossed-aside for now and a total reformulation of the wavefunction ψ, as governed by the Schrödinger wave equation prescribed in Eq. (1), will be carried out instead. Essentially, for the purpose of obtaining the true wavefunction solution ψ, one needs to keep in mind that ψ must attain a plane-wave state, that is, $\psi \sim e^{ikx}$, whenever the limit $\lim \omega \to 0$ is invoked in the Schrödinger wave equation of Eq. (1).

2. Reformulation of the Wavefunction $\psi(z)$

§. *The Wavefunction Ansatz.* The Schrödinger wave equation defined in Eq. (1) can be rewritten in dimensionless form as

$$\frac{d^2\psi(z)}{dz^2} + \left(\varepsilon - \tfrac{1}{4}z^2\right)\psi(z) = 0, \tag{3}$$

where z is made explicit here in the argument of the wavefunction $\psi(z)$ through the following dimensionless quantities

$$x = \tfrac{1}{\sqrt{2}}\,az, \quad a = \sqrt{\frac{\hbar}{m\omega}}, \quad \text{and} \quad \varepsilon = \frac{mEa^2}{\hbar^2} = \frac{mE}{\hbar^2}\frac{\hbar}{m\omega} = \frac{E}{\hbar\omega}.$$

At this point, the complete wavefunction solution to Eq. (3), of the form

$$\psi(z) = e^{ikz} f(z), \tag{4}$$

whose wave-number k satisfies the relation $k^2 = \varepsilon$, is assumed. One will be further relieved to find, after substituting Eq. (4) into Eq. (3), that $f(z)$ satisfies a second-order differential equation:

$$f''(z) + 2ikz\, f'(z) - \tfrac{1}{4} z^2 f(z) = 0, \tag{5}$$

where the primes ' and " represent first- and second-order derivatives, respectively, with respect to the z variable.

The temptation of rearing out a possible solution to $\psi(z)$ from any well-known book on the subject of second-order differential equations, especially bearing a strong resemblance to Eq. (5), should be avoided. Here experience will prevail, since all such efforts, inevitably, lead right-back to the start of Eq. (5), merely because the plane-wave e^{ikz} component will always cancel-out completely, the result of which defeats the very purpose of using the initial form of the solution $\psi(z)$ *spoused* in Eq. (4).

It is clear, if one is to keep the plane-wave e^{ikz} anchored onto the complete wavefunction solution of $\psi(z)$, that another mode of solution for solving the second-order differential equation for $f(z)$ in Eq. (5), whatever that may be, has to be advanced immediately.

§. *The Three-Term Recursion Relation.* Λ suitable expansion for $f(z)$ in Eq. (5) is

$$f(z) = \sum_{n=0}^{\infty} c_n D_{n+\nu}(z) \tag{6}$$

where from **Chapter 8** of Magnus *et al* (1966) [2], $D_{n+\nu}(z)$ is Weber's function, or the parabolic cylinder functions, indexed by $n + \nu$. Here the index ν is a *free-parameter* and will be left arbitrary for now, as its true mathematical scope, as well as physical meaning, will emerge more naturally, respectively, in sections §3 and §4 of the present chapter.

If one inquires into pp. 324 of Magnus *et al* (1966) [2] for the second-order differential equation governing Weber's function $D_{n+\nu}(z)$, one finds

$$D''_{n+\nu}(z) + \left(n + \nu + \tfrac{1}{2} - \tfrac{1}{4}z^2\right)D_{n+\nu}(z) = 0, \qquad (7)$$

with a similar relation for it's first derivative, from pp. 327 of Magnus *et al* (1966) [2],

$$D'_{n+\nu}(z) = \tfrac{1}{2}\left[(n+\nu)D_{n+\nu-1}(z) - D_{n+\nu+1}(z)\right], \qquad (8)$$

where the prime symbols ' and " in Eqs. (7) and (8) denote respectiveely the first- and second-order derivatives of $D_{n+\nu}(z)$ with respect to z. By carefully inserting the series expansion given by Eq. (6) for $f(z)$ into Eq. (5), one can readily use the relations obtained in Eqs. (7) and (8) to get a hold of a *three-term recursion relation* (up to a sign change) for the expansion coefficients c_n of $f(z)$, provided one equates the index $n + \nu$ of Weber's function $D_{n+\nu}(z)$ term by term, like so

$$ik(n + \nu + 1)c_{n+1} - \left(n + \nu + \tfrac{1}{2}\right)c_n - ik\,c_{n-1} = 0, \qquad (9)$$

where i is the pure imaginary complex number, which – unless otherwise stated – will be taken for granted throughout this monograph.

If one is ever to settle past the forewarnings mentioned at the start of this section of the chapter, as well as to have the good fortune of solving the Schrödinger wave equation with the plane-wave component e^{ikz} anchored to the solution $\psi(z)$, which is almost insisted on by Eq. (4), then, there is really no avenue of escape from the three-term recursion relation displayed in Eq. (9). Consequently, confronting this three-term recursion relation has now suddenly become so deeply-seated to the nature of the problem at hand. This is a stunning contrast to the original formulation of the wavefunction for the simple harmonic oscillator as first delivered by Erwin Schrödinger, since in that instance only a two-term recursion relation was ever witnessed (e.g., see Schiff (1965) [1]).

§. *Solution of the Three-Term Recursion Relation.* There is no more doubt now! The three-term recursion relation in Eq. (9) has to be solved exactly or dealt with as precisely as possible before any claim of a complete solution of the wavefunction $\psi(z)$ to the Schrödinger wave equation for the harmonic oscillator can be made.

To commence with, assume that the expansion coefficients c_n in Eq. (9) can be solved by

$$c_n = \frac{(-1)^n (ik)^{-n}}{\Gamma(n+\nu+1)\Gamma\left(\frac{1}{2}-n\right)} a_n, \tag{10}$$

a_n being an auxiliary coefficient to be determined and $\Gamma(x)$ denoting the usual Gamma function, as defined, for example, in **Chapter 1** of Magnus *et al* (1966) [2].

The auxiliary coefficients a_n are determined by inserting the c_n's of Eq. (10) directly into the three-term recursion relation of Eq. (9), since it is anticipated that the resulting recursion formula obtained for the a_n's will – at the very least – be recognized.

One can certainly check that the resulting three-term recursion relation for a_n and its peers $a_{n\pm1}$, after some simplification, give

$$\frac{(-1)^n (ik)^{-n}}{\Gamma(n+\nu+1)\Gamma\left(\frac{3}{2}-n\right)} \left\{ \left(n+\nu+\tfrac{1}{2}\right)\left(\tfrac{1}{2}-n\right)a_n + (n+\nu)k^2 a_{n-1} \right.$$
$$\left. + \left(\tfrac{1}{2}-n\right)\left(-\tfrac{1}{2}-n\right)a_{n+1} \right\} = 0, \tag{11}$$

where the property $x\Gamma(x) = \Gamma(x+1)$ for the Gamma function, from **Chapter 1** of Magnus *et al* (1966) [2], has been applied.

Next, for brevity, introduce

$$v \equiv k^2, \quad \alpha = -n-v, \quad \beta = \tfrac{1}{2}-n \quad \text{and} \quad y = v,$$

so then the three-term recursion relation for the a_n's inside the curly brackets of Eq. (11) can be rewritten more compactly as

$$\beta(1-\beta+y)a_n - \alpha\, y\, a_{n-1} + \beta(\beta-1)a_{n+1} = 0, \tag{12}$$

which is opportune for seizing an obvious connection with *Kummer's first confluent hypergeometric function,* denoted here in the notation of

Chapter 13 of Ambramowitz and Stegun (1964) [3] by $M(\alpha, \beta, y)$. Alternative notations for $M(\alpha, \beta, y)$ is the hypergeometric function $_1F_1(\alpha; \beta; y) = M(\alpha, \beta, y)$, as mentioned also by Ambramowitz and Stegun, pp. 504.

There is no need to hesitate here. One immediately connects a_n in Eq. (12) with the confluent hypergeometric function, that is, $a_n = M(\alpha, \beta, y)$, to realize

$$\beta(1 - \beta + y)M(\alpha, \beta, y) - a\,y\,M(\alpha + 1, \beta + 1, y)$$
$$+ \beta(\beta - 1)M(\alpha - 1, \beta - 1, y) = 0, \qquad (13)$$

which as verified by Eq. 13.4.7 of Ambramowitz and Stegun (1964) [3] is precisely the recursion relation satisfied by $M(\alpha, \beta, y)$. This establishes, quite convincingly, that the confluent hypergeometric function $M(\alpha, \beta, y)$ is an exact solution for the coefficients a_n of the recursion formula defined by Eq. (12).

Although the present author realized the solution specified by Eq. (12) for the a_n coefficients first, there is yet another solution.

To see this explicitly, one substitutes

$$c_n = (ik)^{-n} b_n,$$

and obtains a similar three-term recursion relation for the b_n's:

$$-(ik)^{-n}\left\{\left(n + v + \tfrac{1}{2}\right)b_n - k^2 b_{n-1} - (n + v + 1)b_{n+1}\right\} = 0. \qquad (14)$$

As before, the symbols

$$v \equiv k^2, \quad \alpha = v + \tfrac{1}{2}, \quad \beta = \tfrac{1}{2} - n \quad \text{and} \quad y = -v,$$

can be used to bring Eq. (14) into the form,

$$(\beta - \alpha - 1)b_{n+1} + (1 - \alpha - y)b_n + y\,b_{n-1} = 0, \qquad (15)$$

which is now prone for an encounter with *Kummer's second independent hypergeometric function*, denoted in the notation used on pp. 504 of Ambramowitz and Stegun (1964) [3] by $U(\alpha, \beta, y)$. Another notation for Kummer's second independent hypergeometric function is

$\Psi(\alpha; \beta; y) = U(\alpha; \beta; y)$ as one can see from **Chapter 9.21**, pp. 1058, of Gradshteyn and Rhyzik (1980) [4]. Its formal definition in terms of Kummer's first confluent hypergeometric function, $M(\alpha; \beta; y)$, is more common and, since it will be needed later on, it is shown in Eq.(20) of section §3 of this chapter.

It is a simple matter of verifying whether Kummer's second confluent hypergeometric function is indeed connected with the b_n's in the recursion formula of Eq. (15). Assume then that the relation $b_n = U(\alpha, \beta, y)$ holds in Eq. (15), and observe

$$(\beta - \alpha - 1)U(\alpha, \beta - 1, y) + (1 - \beta - y)U(\alpha, \beta, y) +$$
$$+ y U(\alpha, \beta + 1, y) = 0,$$

which when lined-up against Eq. 13.4.16 of Ambramowitz and Stegun (1964) [3] is in complete agreement with the recursion relation for Kummer's second confluent hypergeometric function $U(\alpha, \beta, y)$.

In short, there are two linearly independent solutions that solve for the expansion coefficients c_n in the recursion formula exactly, both of which are directly connected with Kummer's first and second confluent hypergeometric functions.

Generally, one can place both independent solutions under one common coefficient, say, c_n,

$$c_n = A \frac{(-1)^n (ik)^{-n}}{\Gamma(n + v + 1)\Gamma\left(\frac{1}{2} - n\right)} M\left(-n - v, \tfrac{1}{2} - n, k^2\right) \tag{16}$$
$$+ B(ik)^{-n} U\left(v + \tfrac{1}{2}, \tfrac{1}{2} - n, -k^2\right),$$

where here A and B are arbitrary constants with the original indices restored in the arguments for the confluent hypergeometric functions, noting that $v \equiv k^2$ is present in both the linearly independent terms above.

§. ***Linear Independence and the Wronskian.*** Formal justification that both of the terms specified through the c_n coefficients in Eq. (16) are, indeed, linearly independent will now be verified, first by splitting c_n into two parts, as in $c_n^{(1)}$ and $c_n^{(2)}$, such that:

$$c_n \equiv A c_n^{(1)} + B c_n^{(2)}$$

$$= A \frac{(-1)^n (ik)^{-n}}{\Gamma(n+v+1)\Gamma\left(\frac{1}{2}-n\right)} M\left(-n-v, \tfrac{1}{2}-n, k^2\right) +$$

$$+ B(ik)^{-n} U\left(v+\tfrac{1}{2}, \tfrac{1}{2}-n, -k^2\right), \tag{17}$$

and then forming the determinant (i.e., $\det | \ |$) of $c_n^{(1)}$ and $c_n^{(2)}$, since one only needs to check if the following *Wronskian* condition is met:

$$W_n = \det \begin{vmatrix} c_n^{(1)} & c_n^{(2)} \\ c_{n+1}^{(1)} & c_{n+1}^{(2)} \end{vmatrix} = c_n^{(1)} c_{n+1}^{(2)} - c_{n+1}^{(1)} c_n^{(2)} \neq 0 .$$

This definition of the Wronskian follows the formalism of **Chapter 2** of Bender and Orszag (1978) [5], and is practically analgous to that used in ordinary differential equations, so is just as well placed here in the context of finite-difference equations.

It useful to first recast each of the coefficients, $c_n^{(1)}$ and $c_n^{(2)}$, specified in Eq. (17), into their more compact forms,

$$c_n^{(1)} = \frac{(-1)^n (-ik)^{-n}}{\Gamma(n+v+1)\Gamma\left(\frac{1}{2}-n\right)} M\left(-n-v, \tfrac{1}{2}-n, k^2\right)$$

$$\equiv e^{k^2} \frac{(-1)^n (-ik)^{-n}}{\Gamma(\alpha-\beta+1)\Gamma(\beta)} M\left(\alpha, \beta, -k^2\right),$$

and

$$c_n^{(2)} = (ik)^{-n} U\left(v+\tfrac{1}{2}, \tfrac{1}{2}-n, -k^2\right)$$

$$\equiv (ik)^{-n} U\left(\alpha, \beta, -k^2\right),$$

where the abbreviations, $\alpha = v + \tfrac{1}{2}$, $\beta = \tfrac{1}{2} - n$, together with Kummer's transformation formula $M(\alpha,\beta,y) = e^y M(\beta-\alpha, \beta, -y)$, taken from Eq. 13.1.27 of Ambramowitz and Stegun (1964) [3], to change the sign of the argument of $M(\alpha,\beta,y)$ in the second line of the first expansion coefficient $c_n^{(1)}$ above, are now in use.

Next, from Eqs. 13.4.13 and 13.4.24 of Ambramowitz and Stegun (1964) [3], one can acquire two more properties attached with the first-derivatives of the following confluent hypergeometric functions,

$$M\left(\alpha, \beta - 1, -k^2\right) = M\left(\alpha, \beta, -k^2\right) + \frac{z}{(\beta - 1)} M'\left(\alpha, \beta, -k^2\right),$$

$$U\left(\alpha, \beta - 1, -k^2\right)$$

$$= \frac{(1 - \beta)}{(1 + \alpha - \beta)} U\left(\alpha, \beta, -k^2\right) + \frac{z}{(1 + \alpha - \beta)} U'\left(\alpha, \beta, -k^2\right),$$

where $'$ indicates differentiation with respect to the argument. From here, one can more or less go straight on to compute the Wronskian, to find, after some lenghty algebra,

$$W_n \equiv c_n^{(1)} c_{n+1}^{(2)} - c_{n+1}^{(1)} c_n^{(2)}$$

$$= \frac{(-1)^n (ik)^{-2n-1} e^{k^2}}{\Gamma(1 + \alpha - \beta)\Gamma(\beta)} \left[\frac{2(1 - \beta)}{(1 + \alpha - \beta)} M\left(\alpha, \beta, -k^2\right) U\left(\alpha, \beta, -k^2\right) + \right.$$

$$\left. + \frac{k^2}{(1 + \alpha - \beta)} W\{y_1, y_2\} \right]$$

$$= \frac{(-1)^n (ik)^{-2n-1} e^{k^2}}{\Gamma(1 + \alpha - \beta)\Gamma(\beta)} \left[\frac{2(1 - \beta)}{(1 + \alpha - \beta)} M\left(\alpha, \beta, -k^2\right) U\left(\alpha, \beta, -k^2\right) + \right.$$

$$\left. + \frac{k^{n + \frac{1}{2}} e^{-k^2}}{(1 + \alpha - \beta)} \frac{\Gamma(\beta)}{\Gamma(\alpha)} \right], \quad (18)$$

where it is worthwhile to note that Eq. 13.1.22 of Ambramowitz and Stegun (1964) [3] can be utilized efficiently since it allows the Wronskian, $W\{y_1, y_2\} = \left(-k^2\right)^{-\beta} e^{k^2} \Gamma(\alpha)/\Gamma(\beta)$, for the confluent hypergeometric functions, $y_1 = M\left(\alpha, \beta, -k^2\right)$ and $y_2 = U\left(\alpha, \beta, -k^2\right)$, to be used in the last line of Eq. (18). This is very comforting to know, because it also shows that the linear independence of $c_n^{(1)}$ and $c_n^{(2)}$ is mutual throughout, on account of the fact that their Wonskian will never vanish. Consequently, the criterion of a non-zero determinant in the computation of the Wronskian, $W_n \equiv c_n^{(1)} c_{n+1}^{(2)} - c_{n+1}^{(1)} c_n^{(2)}$, has now been more than satisfactorily fulfilled.

The conclusion reached in Eq. (18) of a non-zero Wronskian for $c_n^{(1)}$ and $c_n^{(2)}$ is, of course, as good as a guarantee that the complete wavefunction $\psi(z)$ can also be expressible as a sum of two linearly independent functions. This result is taken advantage of in the next section below for completing the formulation of the wavefunction.

3.   The Complete Wavefunction $\psi(z)$

§. *Relation with the Laguerre Functions.* Rather than wind-down an infinite series involving the confluent hypergeometric functions, $M\left(-n-v, \frac{1}{2}-n, k^2\right)$ and $U\left(v+\frac{1}{2}, \frac{1}{2}-n, -k^2\right)$, whose indices from a quantum computing point of view will only clutter the organization of the terms – inevitably true of any type of series expansion of the wavefunction $\psi(z)$ – it will be found more convenient, as well as conventional, to redefine the confluent hypergeometric functions in terms of the Laguerre functions $L_v^\alpha(x)$.

With the aid of the following definition of the Laguerre functions $L_v^\alpha(x)$, from Eq. 9.50 of Andrews (1985) [6][*],

$$L_v^\alpha(x) = \frac{\Gamma(v+\alpha+1)}{\Gamma(v+1)\Gamma(\alpha+1)} M(-v, \alpha+1, x), \qquad (19)$$

one rapidly learns, indeed, that when the index or the order of a known orthogonal polynomial is no longer an integer, such polynomials no longer possess finite terms but now are an infinite series. In this sense, the associated Laguerre polynomials become the Laguerre functions shown in Eq. (19) of arbitrary index v. A list of

---

[*] Note that Andrews uses semi-colons in the arguments of the confluent hypergeometric functions, whereas the present monograph maintains commas as the standard to be consistent with Ambramowitz and Stegun (1964) [3].

other orthogonal polynomials known to be generalized this way can also be found in **Chapter 9** of Andrews (1985) [6].

The transformation defined in Eq. (19) will also prove to be profitable for Kummer's second independent hypergeometric function, whose formal definition is quoted below from Eq. 13.1.3 of Ambramowitz and Stegun (1964) [3],

$$U(a, b, z) =$$

$$= \frac{\pi}{\sin \pi b}\left\{\frac{M(a, b, z)}{\Gamma(1+a-b)\Gamma(b)} - z^{1-b}\frac{M(1+a-b, 2-b, z)}{\Gamma(a)\Gamma(2-b)}\right\}. \tag{20}$$

Now it's time to piece together the *complete* wave function as follows. First, insert the result obtained in Eq. (16) for the expansion coefficients, c_n, directly into the wavefunction $\psi(z)$ ansatz of Eq. (4) (or Eq. (6)). Next, use the defining relation for the Laguerre functions in Eq. (19), together with the formal definition of $U(a, b, z)$ given by Eq. (20). The final result, then, for $\psi(z)$, as shown below, is quite breathtaking, so it rightly deserves to secure its own space $\left(v = k^2\right)$:

$$\psi(z) \equiv A\,e^{ikz}\,y_1 + B\,e^{ikz}\,y_2,$$

$$= A\,e^{ikz}\sum_{n=0}^{\infty}\frac{(-1)^n (ik)^{-n} M\left(-n-v, \frac{1}{2}-n, k^2\right)}{\Gamma(n+v+1)\Gamma\left(\frac{1}{2}-n\right)}D_{n+v}(z) +$$

$$+ \; B\,e^{ikz}\sum_{n=0}^{\infty}(ik)^{-n}U\left(v+\tfrac{1}{2}, \tfrac{1}{2}-n, -k^2\right)D_{n+v}(z), \tag{21}$$

$$= A\,e^{ikz}\sum_{n=0}^{\infty}\frac{(-1)^n (ik)^{-n} L_{n+v}^{-\frac{1}{2}-n}\left(k^2\right)}{\Gamma\left(v+\frac{1}{2}\right)}D_{n+v}(z) +$$

$$+ B\pi\,e^{ikz}\sum_{n=0}^{\infty}(-1)^n (ik)^{-n} e^{-k^2}\left[\frac{L_{n+v}^{-\frac{1}{2}-n}\left(k^2\right)}{\Gamma\left(v+\frac{1}{2}\right)} - \left(k^2\right)^{n+\frac{1}{2}}\frac{L_{v-\frac{1}{2}}^{n+\frac{1}{2}}\left(k^2\right)}{\Gamma(n+v+1)}\right]D_{n+v}(z).$$

§. *The Physical Solution.* Eq. (21) shows that the complete wavefunction $\psi(z)$ comprises of a product of a plane-wave solution,

e^{ikz}, and a series solution. In reality, there are two independent terms for the series solution part of $\psi(z)$, as a consequence of the connection drawn out in the previous section with Kummer's two linearly independent confluent hypergeometric functions. Here the product form of the wavefunction $\psi(z)$ that one sees directly in Eq. (21), makes it ideal to refer to the confluent hypergeometric functions respectively as the two physical solutions, $y_{1,2}$, of $\psi(z)$.

Therefore, the complete wavefunction $\psi(z)$ of Eq. (21), forgetting about the arbitrary constants in that equation, has the augmented form

$$\psi(z) \sim e^{ikz}\, y_{1,2},$$

which, incidentally, stresses the highly unusual nature of plane-waves travelling in the same direction of a region of space. One is reminded here that for a function to possess linearly independent solutions of the type, $e^{ikz} y_{1,2}$, that is, whose exponential arguments possess the same sign, is, of course, considered to be of an "unknown rarity" in quantum physics, or even in pure mathematics – surely.

But before anyone dares to pass a doubt over the rarity nature of $e^{ikz} y_{1,2}$, it deserves to be mentioned, once more, that the computation of the Wronskian for the expansion coefficients $c_n^{(1)}$ and $c_n^{(2)}$ in section §2 proved indispensable, since there it was indicated that the linear independence of $c_n^{(1)}$ and $c_n^{(2)}$, implies the mutual independence of both y_1 and y_2 to hold strongly.

If further convincing is required, one can always turn to the choice of using Weber's function in the wavefunction ansatz of Eq. (4), since one realizes that both real and complex solutions in the form

$$D_{n+v}(z),\ \ D_{n+v}(-z),\ \ D_{-n-v-1}(iz) \quad \text{and} \quad D_{-n-v-1}(-iz),$$

cover all the possible types of parabolic cylinder functions that are considered equally exceptional to use in formulating a complete wavefunction solution $\psi(z)$ to Eq. (3). In fact, the reader can refer to the bottom of pp. 324 of Magnus *et al* (1966) [2] to verify, that each of the parabolic cylinder functions outstretched above solve precisely the same second-order differential equation – namely, Weber's equation. In the present manuscript, Weber's equation is no more different than the Schrödinger's equation specified (in dimensionless form) in Eq. (3), with the index ε replaced by $n + v + \frac{1}{2}$.

Combined with the linear independence property mentioned in the preceding paragraphs, one has now been exposed to a total of 8 classes of possible solutions for representing the complete wavefunction $\psi(z)$ in Eq. (21). From a quantum mechanical standpoint, one can dismiss $D_{-n-v-1}(iz)$ and $D_{-n-v-1}(-iz)$ amongst the classes of solutions, since physically acceptable solutions for the Schrödinger wave equation in quantum mechanics must adhere to the requirement of being of the square-integrable L^2 class. That is, $\psi(z)$ must vanish at the boundaries, i.e., $\psi(z) \to 0$ with respect to the asymptotic limit, $|z| \to \infty$.

One can further insure $D_{n+v}(z)$ and $D_{n+v}(-z)$ as the only two physically acceptable classes of solutions to $\psi(z)$ in Eq. (19) by casting an "eyebrow" onto the asymptotic form of Weber's function,

$$D_{n+v}(\pm z) \approx e^{-\frac{1}{4}z^2} (\pm z)^{n+v} \left\{ 1 - \frac{(n+v)(n+v-1)}{2z^2} + \cdots\cdots \right\},$$

$$|\arg z| < 3\pi/4$$

as adapted from section §8.1.6 of Magnus *et al* (1966) [2]. This provides a formal justification for employing $D_{n+v}(z)$ in the expansion for $f(z)$ in Eq. (6), because the plane-wave e^{ikz} solution part of $\psi(z)$, as seen from the asymptotic form of $D_{n+v}(z)$ displayed above – evidently – will always be under the domination of the exponential term, $e^{-\frac{1}{4}z^2}$, which itself has a permanent presence in the physical solutions, $y_{1,2}$, part of $\psi(z)$.

4. The Eigenvalue Search

Determination of the *discrete eigenvalues* of the wavefunction solution to the Schrödinger wave equation in quantum mechanics, requires the ability to forecast the behaviour of the wavefunction $\psi(z)$ for large values of its argument; that is, in the limit $z \to \infty$, which is called the *asymptotic region of space*. A comprehensible analysis is elaborated in **Chapter 9** of Morse & Feschbach (1953) [7], with emphasis on the variational principle as a highly accurate method of calculation of the eigenvalues.

Naturally, the objective of analysing the asymptotic content of the wavefunction, using various successive approximation schemes or perturbation techniques as well as variational methods, is one of the highlights of that reference, since the techniques employed there by Morse and Feschbach are designed to culminate the domineering power behind the method of *asymptotic analysis*. This is considered particularly vital for determining whether or not the behaviour of the wavefunction in the asymptotic region is strictly convergent or divergent. That is, whether $\psi(z)$ has an exponential decaying or increasing behaviour respectively with its argument, z, or if it varies, instead, as some power-law in z. There are, of course, certain preparations, as shall drawn-out below, that have to be made in order to reach this goal.

§. *Toward a Formal Expansion of* $\psi(z)$. First, it makes perfect sense to seek a formal power-series representation for the complete wavefunction, because one can almost "peer down" at the wavefunction directly, i.e., identify the trailing terms of the series, to validate its true asymptotic form or behaviour. In practice, however, one begins with an analysis of the *expansion coefficients*, say a_n, of the wavefunction, whose power-series solution form, taken for the moment to hold arbitrary weight $w(z)$, is assumed formally to be represented as

$$\psi(z) = w(z)\sum a_n z^n,$$

where the growth of $\psi(z)$ is to be measured directly by evaluating the ratio of its expansion coefficients, a_n, with the large limit $n \to \infty$

imposed (i.e., $\lim n \to \infty \, a_{n+1}/a_n$). As suggested, evaluation of the ratio of the expansion coefficients in the wavefunction, or simply the ratio test, of the wavefunction $\psi(z)$, is a strongly reliable procedure for the accurate determination of the discrete eigenvalues of the Schrödinger wave equation (Morse & Feschbach (1953) [7]).

Before proceeding, however, be aware that it is an almost impossible challenge to fully close the summation over the index n in the wavefunction solution of Eq. (21), so arriving at a known transcendental form of the wavefunction $\psi(z)$, in this case, still remains elusive. This means that the chances of "spotting" any of the eigenvalues of the wavefunction solution of Eq. (21) directly is also somewhat diminished as a result.

For this reason, one must look for an alternative way to *regroup* the infinite sums in the power-series expansion for $\psi(z)$, in the hope that there exists an expansion that one can consider more expedient to the task of analysing the expansion coefficients of $\psi(z)$ more easily.

Fortunately, the present author has managed to derive an alternate, yet, extremely useful expansion for $D_{n+v}(z)$, which avoids a a straight-out expansion of the Laguerre and parabolic cylinder functions in the complete wavefunction $\psi(z)$ of Eq. (21). The reader should be convinced that the more direct tactic of a straight-out expansion in the complete wavefunction $\psi(z)$ of Eq. (21) only renders the expansion coefficients of $\psi(z)$ to become even more intractable for the actual determination of the eigenvalues. Therefore, the following new expansion of $D_{n+v}(z)$ below is intended to prepare the wavefunction $\psi(z)$ for a more complete asymptotic analysis at a future date.

§. ***An Expansion for*** $\boldsymbol{D_{n+v}(z)}$**.** It is high-time to pause here and find out the significance of the free-parameter v introduced in section §2. First, recall from that section, that the parameter v was carefully equated to the square of the wave-number, according to the prescription $v \equiv k^2$, where k is the wave-number, $k = \sqrt{E/\hbar\omega}$. Recall also that the energy E is itself interchangeable with its dimensionless quantity ε (defined below Eq. (3) in the text). There is, of course, no question about the real-valuedness of the energy eig-

15

envalues, E, since this is an immediate consequence of the well-known Hermitian or self-adjoint operator properties attached with the Schrödinger wave equation in quantum mechanics (e.g., see Flugge (1974) [8]). In addition, one confirms that the total energy, E or ε, for the quantum mechanical oscillator as being necessarily non-negative, i.e., $E \geq 0$ $\left(\text{or } \varepsilon \geq 0\right)$, because the potential energy of the oscillator given by the second-term on the left-hand-side of the Schrödinger wave equation (defined by Eq. (1)) is itself a non-negative function. Accordingly, this singles out the fact that the energy eigenvalues ε (or E) will be strictly non-negative as well as real-valued, so that v is inherently a positive real-valued quantity.

The points raised above, only justifies, once more, the decision to use $D_{n+v}(z)$ on more solid physical grounds. Such matters, as will be seen shortly below, also prevent the desired the expansion of $D_{n+v}(z)$ from being contested further; though, the reader should be aware that other expansions of Weber's function do exist, such as those obtained by Erdelyi (1953) [9] (see also Daniel (2001) [10]), and are available to be compared with the result derived below in Eq. (24).

The desired expansion of $D_{n+v}(z)$ begins first with the integral representation of Weber's function,

$$D_p(z) = \frac{2^{p+\frac{1}{2}}}{\sqrt{\pi}} e^{-\frac{\pi}{2}pi} e^{\frac{z^2}{4}} \int_{-\infty}^{\infty} x^p e^{-2x^2+2ixz}\, dx, \qquad \left(\text{Re } p > -1\right),$$

from Eq. 9.241, pp. 924 of Gradshteyn and Rhyzik (1980) [9], and secondly with the generating function for the Hermite polynomials,

$$\exp\left(2xz - z^2\right) = \sum_{n=0}^{\infty} \frac{H_n(x)}{n!}\, z^n ,$$

from pp. 252, Magnus *et al* (1966) [2].

Accordingly,

$$D_{n+v}(z) = \frac{2^{n+v+\frac{1}{2}}}{\sqrt{\pi}} e^{-i\frac{\pi}{2}(n+v)} e^{\frac{1}{4}z^2} \int_{-\infty}^{\infty} t^{n+v} e^{-2t^2+2itz}\, dt, \qquad \left(\text{Re } v > -1\right), \qquad (22)$$

$$= \frac{2^{n+v+\frac{1}{2}}}{\sqrt{\pi}} e^{-i\frac{\pi}{2}(n+v)} e^{-\frac{1}{4}z^2} \int\limits_{-\infty}^{\infty} t^{n+v} e^{-2t^2} \sum_{n=0}^{\infty} \frac{1}{\ell!} \left(\frac{iz}{\sqrt{2}}\right)^{\ell} H_{\ell}\left(\sqrt{2}t\right) dt$$

$$= \frac{2^{n+v+\frac{1}{2}}}{\sqrt{\pi}} e^{-i\frac{\pi}{2}(n+v)} e^{-\frac{1}{4}z^2} \sum_{n=0}^{\infty} \frac{1}{\ell!} \left(\frac{iz}{\sqrt{2}}\right)^{\ell} \int\limits_{-\infty}^{\infty} t^{n+v} e^{-2t^2} H_{\ell}\left(\sqrt{2}t\right) dt$$

$$= \frac{2^{\frac{n+v}{2}}}{\sqrt{4\pi}} e^{-\frac{1}{4}z^2} \sum_{n=0}^{\infty} \frac{1}{\ell!} \left(\frac{iz}{\sqrt{2}}\right)^{\ell} e^{\frac{i\pi\ell}{2}} \cos \frac{\pi}{2}(n+v+\ell) \int\limits_{-\infty}^{\infty} e^{-u} u^{\frac{n+v-1}{2}} H_{\ell}\left(\sqrt{u}\right) du$$

The integral on the last line of Eq. (22) above "stares" out a known
association with the hypergeometric function $_2F_1(a,b;c;x)$ as defined
in Eq. 7.383(4), pp. 839, of Gradshteyn & Ryhzik (1980) [4],

$$\int\limits_0^{\infty} x^{a-\frac{1}{2}n-1} e^{-bx} H_n\left(\sqrt{x}\right) dx = 2^{\frac{n}{2}} \Gamma(a) b^{-a} \,_2F_1\left(-\tfrac{1}{2}n, \tfrac{1}{2}(1-n); 1-a; b\right),$$

and, in turn, is akin also to the Gegenbauer polynomial $C_n^{\lambda}(x)$ in
Chapter 5, pp. 220, of Magnus *et al* (1966) [2],

$$C_n^{\lambda}(x) = 2^n \frac{(\lambda)_n}{n!} x^n \,_2F_1\left(-\tfrac{1}{2}n, \tfrac{1}{2}(1-n); -n+\lambda+1; x^{-2}\right),$$

which, when employed together, permit for a swift evaluation of the
integral on the right-hand side of the last line in Eq. (22). This leads
to the result

$$D_{n+v}(z) = \frac{2^{\frac{n+v}{2}}}{\sqrt{4\pi}} e^{-\frac{1}{4}z^2} \sum_{n=0}^{\infty} \frac{\left(iz/\sqrt{2}\right)^{\ell}}{\ell!} \frac{2^{\frac{\ell}{2}} e^{\frac{i\pi\ell}{2}}}{\ell!} \cos \frac{\pi}{2}(n+v+\ell) \times$$

$$\times\ \Gamma\!\left(\frac{n+v+\ell+1}{2}\right) {}_2F_1\left(-\tfrac{1}{2}\ell, \tfrac{1}{2}(1-\ell); \tfrac{1}{2}(1-n-v-\ell); 1\right)$$

$$= \frac{2^{\frac{n+v}{2}}}{\sqrt{4\pi}} e^{-\frac{1}{4}z^2} \sum_{n=0}^{\infty} \frac{\left(iz/\sqrt{2}\right)^{\ell}}{\ell!} \frac{2^{\frac{\ell}{2}} e^{\frac{i\pi\ell}{2}}}{\ell!} \cos \frac{\pi}{2}(n+v+\ell) \times$$

$$\times\ \frac{2^{-\ell} \ell!}{\left(\frac{1+n+v-\ell}{2}\right)_{\ell}} \Gamma\!\left(\frac{n+v+\ell+1}{2}\right) C_{\ell}^{\frac{1}{2}(n+v+1-\ell)}(1) \qquad (23)$$

where $(a)_{\ell} = \frac{\Gamma(a+\ell)}{\Gamma(a)}$ is a Pochammer symbol, as frequently practiced to represent factorial type arguments; for example, in **Chapter 1** of Magnus *et al* (1966) [2].

The expansion reached in Eq. (23) for $D_{n+\nu}(z)$ can, of course, be freed from its connection with the hypergeometric function or its close-by polynomial relative. For instance, simply replace $C_n^\lambda(1)$ in the last line of Eq. (23) with its binomial equivalent, i.e., the value of the Gegenbauer polynomial $C_n^\lambda(x)$ at $x = 1$. This value can be easily ascertained from Table 22.4, pp. 777 of Ambramowitz and Stegun (1964) [3],

$$C_n^\lambda(1) = \binom{n + 2\lambda - 1}{n} \qquad (\lambda \neq 0),$$

recalling that $\binom{a}{b} = \frac{\Gamma(a+1)}{\Gamma(a-b+1)\,\Gamma(b+1)}$ represents a typical binomial coefficient.

Gathering the last result in the previous paragraph, one finally expresses $D_{n+\nu}(z)$ as

$$D_{n+\nu}(z) = \frac{2^{\frac{n+\nu}{2}}}{\sqrt{4\pi}} e^{-\frac{1}{4}z^2} \sum_{n=0}^{\infty} \frac{1}{\ell!} \frac{(-z)^\ell}{\ell!} \cos\tfrac{\pi}{2}(n+\nu+\ell)\, \Gamma\!\left(\frac{n+\nu+\ell+1}{2}\right) \times$$

$$\times \ \frac{2^{-\ell}\ell!}{\left(\frac{1+n+\nu-\ell}{2}\right)_\ell} \binom{n+\nu}{\ell}, \tag{24}$$

which completes the desired expansion for $D_{n+\nu}(z)$. Note, to avoid concealing the exponential character attached with this expansion, no further attempt to cancel out the $\ell!$ in Eq. (24) is made at this point.

§. *The Power Series of $\psi(z)$.*

Primed with the expansion of $D_{n+\nu}(z)$ in Eq. (24), it is an elementary step-away to obtain a power-series expansion of the wavefunction $\psi(z)$. Simply, insert $D_{n+\nu}(z)$ from Eq. (24) into the right hand side of the wavefunction solution $\psi(z)$ of Eq. (19), then reverse the order of summation in the indices n and ℓ. The result is a power-series for $\psi(z)$,

$$\psi(z) = A e^{ikz}\, y_1 + B e^{ikz}\, y_2, \tag{25}$$

$$= A\, e^{ikz}\, e^{-\frac{1}{4}z^2} \sum_{\ell=0}^{\infty} A_\ell z^\ell + B\, e^{ikz}\, e^{-\frac{1}{4}z^2} \sum_{\ell=0}^{\infty} B_\ell z^\ell,$$

where, in order to simplify the exact expressions for the expansion coefficients of the power-series of $\psi(z)$, the $\ell!$ arguments are now eliminated to arrive at

$$A_\ell = \frac{2^{\frac{v}{2}}}{\sqrt{4\pi}}\, 2^{-\ell}\, (-1)^\ell \sum_{n=0}^{\infty} (-1)^n \binom{n+v}{\ell} \frac{\Gamma\!\left(\frac{n+v+\ell+1}{2}\right)}{\left(\frac{1+n+v-\ell}{2}\right)_\ell} \left(\frac{\sqrt{2}}{ik}\right)^n \times$$

$$\times\ \cos\tfrac{\pi}{2}(n+v+\ell)\, \frac{L_{n+v}^{-\frac{1}{2}-n}\!\left(k^2\right)}{\Gamma\!\left(v+\frac{1}{2}\right)},$$

and

$$B_\ell = 2^{\frac{v}{2}} \sqrt{\pi}\, 2^{-\ell}\, (-1)^\ell \sum_{n=0}^{\infty} (-1)^n \binom{n+v}{\ell} \frac{\Gamma\!\left(\frac{n+v+\ell+1}{2}\right)}{\left(\frac{1+n+v-\ell}{2}\right)_\ell} \left(\frac{\sqrt{2}}{ik}\right)^n \times$$

$$\times\ \cos\tfrac{\pi}{2}(n+v+\ell)\, e^{-k^2} \left[\frac{L_{n+v}^{-\frac{1}{2}-n}\!\left(k^2\right)}{\Gamma\!\left(v+\frac{1}{2}\right)} - \left(-k^2\right)^{n+\frac{1}{2}} \frac{L_{v-\frac{1}{2}}^{\frac{1}{2}+n}\!\left(k^2\right)}{\Gamma\!\left(n+v+1\right)} \right].$$

The form of the expansion coefficients, A_ℓ and B_ℓ, that one notices here, are the most suitably condensed expressions to date, since it will pay to realize that the accuracy of calculating the eigenvalues via a formal power-series expansion of the wavefunction $\psi(z)$, depend critically on "pinning" down the dependence of a_{n+1}/a_n on n in the large limit as $n \to \infty$. The success of this approach depend, undoubtedly, on how one can draw-out the precise mathematical law-like dependence of the wavefunction $\psi(z)$ with its argument z in the asymptotic region of space, which is why analytical expansions such that above is worth the pursuit.

Of course, legislating a mathematical law-like dependence out of the wavefunction $\psi(z)$ via asymptotic analysis of the expansion coefficients of the wavefunction, should be remembered here as being totally analogous to the method outlined in pp. 62 of Schiff

(1955) [1] for determining the eigenvalues of the linear harmonic oscillator.

In the present case, one simply exploits the expressions displayed above for A_ℓ and B_ℓ, in the form of the ratios

$$\lim_{\ell \to \infty} \frac{A_{\ell+1}}{A_\ell} \quad \text{and} \quad \lim_{\ell \to \infty} \frac{B_{\ell+1}}{B_\ell}. \tag{26}$$

Ideally, the asymptotic limit, $\ell \to \infty$, is invoked here purposefully to saturate out the asymptotic behaviour of the wavefunction $\psi(z)$, which, with the aid of the above expressions for A_ℓ and B_ℓ, displayed in the text below Eq. (25), can provide for a highly accurate calculation of the eigenvalues of the wavefunction $\psi(z)$.

It is worth taking another look at the expressions obtained in this section for A_ℓ and B_ℓ. For instance, if one were to head in the opposite direction and solve for the wavefunction $\psi(z)$ using a formal power-series in z, one quickly finds instead a four-term recursion[†] formula to confront with. In other words, as indirect as it may appear, the present work has actually solved a four-term recursion formula satisfied by the expressions above for A_ℓ and B_ℓ. In a manner of speaking, it beyond anyone's guess that using a formal power-series expansion directly first would lead to the expressions given by A_ℓ and B_ℓ, which themselves satisfy a four-term recursion formula.

But for now, Eq. (26) represents the asymptotic analysis part of the *Eigenvalue Search* of the wavefunction $\psi(z)$, starting from its formal power-series expansion given in Eq. (25), which will be taken up in more depth at a later date.

5. Discussion and Final Comments

In the course of the present work on the Schrodinger wave equation, Eq. (1), not only has a full recovery of the plane-wave components of the wavefunction solution $\psi(z)$ for the quantum mechanical oscillator

---

[†] It can be easily shown that the corresponding four-term recursion formula is
$(n+2)(n+1)a_{n+2} + 2ika_{n+1} - (n+\frac{1}{2})a_n - ika_{n-1} = 0$.

been performed but, surprisingly, 8 classes of solutions in total were uncovered in section §3, of which only four of them were considered to be physically acceptable.

However, as quickly learnt in section §3, it is a formidable challenge, indeed, to disprove Erwin Schrödinger original claim of having found the exact solution for the simple harmonic oscillator, since one is immediately faced with a recursion relation of the three-term kind displayed in Eq. (9), especially knowing that the original formulation on the same problem never demanded a recursion relation involving more than two-terms.

While Erin Schrödinger's solution $\psi(z)$ to the linear harmonic oscillator problem is well-known to be celebrated by the Hermite polynomials, the present formulation, on the other hand, has "flawlessly" proven that the true wavefunction solution $\psi(z)$ necessarily involves a linear combination of a product of a plane-wave component coupled with a physical component described by Weber's parabolic cylinder functions.

Undisputably, one of the great triumphs of the results presented in either of Eqs. (21) or (25), is that the complete wavefunction $\psi(z)$ for the linear harmonic oscillator is entirely expressible as products of Weber's function $D_{n+v}(z)$ and the Laguerre functions $L_v^{\alpha}(x)$, which for completeness is flaunted here once more, using the original setting $\left(v \equiv k^2\right)$ and variables of Eq. (1),

$$\psi(x) =$$

$$= A\, e^{i\sqrt{\frac{2mE}{\hbar^2}}\,x} \sum_{n=0}^{\infty} \frac{(-1)^n (ik)^{-n}\, L_{n+v}^{-\frac{1}{2}-n}\left(k^2\right)}{\Gamma\left(v+\frac{1}{2}\right)} D_{n+v}\left(\sqrt{\frac{2m\omega}{\hbar}}\,x\right) +$$

$$+ B\,\pi e^{i\sqrt{\frac{2mE}{\hbar^2}}\,x} \sum_{n=0}^{\infty} (-1)^n (ik)^{-n} e^{-k^2}\left[\frac{L_{n+v}^{-\frac{1}{2}-n}\left(k^2\right)}{\Gamma\left(v+\frac{1}{2}\right)} - \left(-k^2\right)^{n+\frac{1}{2}}\frac{L_{v-\frac{1}{2}}^{n+\frac{1}{2}}\left(k^2\right)}{\Gamma(n+v+1)}\right] \times$$

$$\times\ D_{n+v}\left(\sqrt{\frac{2m\omega}{\hbar}}\,x\right)$$

$$= A\, e^{i\sqrt{\frac{2mE}{\hbar^2}}\,x} e^{-\frac{m\omega}{2\hbar}x^2} \sum_{\ell=0}^{\infty} A_\ell \left(\sqrt{\frac{2m\omega}{\hbar}}\,x\right)^{\ell} + B\, e^{i\sqrt{\frac{2mE}{\hbar^2}}\,x} e^{-\frac{m\omega}{2\hbar}x^2} \sum_{\ell=0}^{\infty} B_\ell \left(\sqrt{\frac{2m\omega}{\hbar}}\,x\right)^{\ell},$$

where here A_ℓ and B_ℓ are the expansions coefficients whose expressions, as mentioned earlier, are given in analytical form below Eq. (25) in the text, with v being the free-parameter discussed at some length in section §2 with respect to having inherited the same magnitude as the square of the wave-number, $k^2 \equiv \varepsilon = 2E/\hbar\omega$. This makes it exclusively clear, that $v = k^2$ holds the same status as the eigenvalue in practical calculations.

Before concluding, there is still one technical reminder in connection with the wave-number, which is contained in the argument of the plane-wave components on the right-hand side of $\psi(z)$, arising from the positive square-root of $k^2 \equiv \varepsilon = 2E/\hbar\omega$. So, strictly speaking, $k = \pm\sqrt{\varepsilon}$. In actual fact then, there is a kind of "accidental degeneracy", quite similar to the degeneracy that one observes in the *angular momenta* of the hydrogen atom in quantum mechanics (e.g., see pp. 86 of Schiff (1953) [1]). This means that the sign of the argument in the plane-wave components of $\psi(z)$ can be opposite in sign to that shown above for $\psi(z)$, so that plane-wave components to $\psi(z)$ "galloping" in the opposite direction are realistically possible as well. Accordingly, the total classes of solution actually increase two-fold in number to that quoted in section §3, i.e., there are 16 exact solutions for the quantum mechanical oscillator.

Nevertheless, it is definite that the missing solutions to Schrödinger's equation for the quantum mechanical oscillator, governed by Eq. (1), have convincingly been uncovered as well as proven to exist. All that remains then, is the determination of the eigenvalues which makes it ever more tempting at this moment to throw out the familiar catch-phrase that Schrödinger's "cat is out of the bag", but it would be better to avoid this cliché and remark instead that Schrödinger's cat has not even been quantized yet.

Following this remark, the present author is optimistic, however, that with the wavefunction of the quantum mechanical oscillator completely formulated, in spite of its inextricable appearance, it is its exact form that will actually spur on the efforts of other researchers and scientists in the overall challenge and pursuit to mine out its eigenvalues.

REFERENCES

1. L. I. Schiff, "Quantum Mechanics," McGraw-Hill, New York (1955).

2. W. Magnus, F. Oberhettinger and R. P. Soni, "Formulas and Theorems for the Special Functions of Mathematical Physics," (3rd Ed.), Springer, Berlin (1966).

3. M. Abramowitz and I. Stegun, "Handbook of Mathematical Functions: with Formulas, Graphs, and Mathematical Tables," Washington, U.S. Govt. Print. Off. (1964).

4. I. S. Gradshteyn and I. M. Ryzhik, "Table of Integrals, Series, and Products," Academic Press, New York (1980).

5. C. M. Bender and S. A. Orszag, "Advanced Mathematical Methods for Scientists and Engineers", McGraw-Hill, New York (1978).

6. L. C. Andrews, "Special Functions for Engineers and Applied Mathematicians," Macmillan, New York (1985).

7. P. M. Morse and H. Feshbach, "Methods of Theoretical Physics II," McGraw-Hill, New York (1953).

8. S. Flügge, "Practical Quantum Mechanics," Springer-Verlag, Berlin (1974).

9. A. Erdelyi, "Higher transcendental functions, based, in part, on notes left by Harry Bateman / and comp. by the staff of the Bateman Manuscript Project," Vol. 2, McGraw-Hill, New York (1953).

10. D. J. Daniel, "Orthogonal Representation of Weber's Function Using Hermite Polynomials," *Journal of Approximation Theory* 113, 156-163 (2001).

CHAPTER 2

The Unique Solution to the Kolmogorov Master Equations for Birth-and-Death

Prior to the author's publication in 2004, the *Kolmogorov equations for birth-and-death* processes, referred to more distinctively in the present chapter as the *Kolmogorov Master equations*, were always thought to have possessed an exact solution. As revealed in the author's publication, the form of the exact solution for this type of Master equation is, surprisingly, found to be truly dependent on the reciprocal of the population number index, n. This means that the population statistics, as well as any related probabilities, attached with *birth-and-death* processes, will almost certainly taper according to this reciprocal dependence on the population number n – contrary to popular belief!

The present chapter begins with a historical discussion of the Kolmogorov equations for birth-and-death, and then builds up to the the recent uncovering of a hidden symmetry inherent in the *Kolmogorov Master equations*. Moreover, as demonstrated for the first time by the present author, the unique solution for the Kolmogorov Master equations must specifically satisfy the initial condition of a Poisson process, $\delta_{n,0}$, which itself represents an unusually daring feat to accomplish in the field of statistical physics.

1. Introduction: A Brief History

There is no need to be in-depth any further than necessary here, than to say, out of all the various types of Markov processes in probability theory, the one type of Markov chain likely to be studied most, is a *linear* "birth-and-death" process. For this process, the transition probabilities for a random variable $X(t)$, denoted here purely out of formality by

$$p_n(t) = \mathscr{P}\{X(t) = n\}, \qquad n = 0,1,2,\ldots \tag{1}$$

follows the assumptions that: at time t the transition probability to

24

jump from the state $n \rightarrow n+1$ in the time interval $(t, t+\Delta t)$ is $\gamma_n \Delta t$, whereas the probability for the transition $n \rightarrow n-1$ in the interval $(t, t+\Delta t)$ is $\alpha_n \Delta t$. Lastly, one assigns $1-\beta_n \Delta t$ for the probability that no transitions from the state n will occur. Under these assumptions, assuming that the probability for all other neighbouring-type transitions to be at least of the order $o(\Delta t)$, i.e., highly improbable, one can then proceed, as in Bailey (1990) [1], to elucidate the entire birth-and-death process as a Taylor series,

$$p_n(t+\Delta t) = \alpha_{n-1} p_{n-1}(t)\Delta t + (1 - \beta_n \Delta t) p_n(t) +$$
$$+ \gamma_{n+1} p_{n+1}(t)\Delta t + o(\Delta t). \tag{2}$$

Since the probability coefficients, α_n, β_n and γ_n, in this relation are generally arbitrary functions of both the index n and the time t, one anticipates a great variety of birth-and-death processes to emerge from Eq. (2) alone.

In the case of linear birth-and-death processes, the linear dependence of the probability coefficients α_n, β_n and γ_n with respect to n is specified but their dependence on t is still arbitrary, and so retain their character as continuous-time functions. More precisely, one assumes

$$\alpha_n = n\alpha(t), \quad \beta_n = -n\beta(t), \quad \text{and} \quad \gamma_n = n\gamma(t). \tag{3}$$

If one now takes the limit $\Delta t \rightarrow 0$ in Eq. (2) and insert also the definitions above for the coefficients, α_n, β_n, and γ_n, one immediately arrives at the birth-and-death equations:

$$\frac{dp_n(t)}{dt} = (n-1)\alpha(t)p_{n-1}(t) + n\beta(t)p_n(t) + (n+1)\gamma(t)p_{n+1}(t), \quad (n \geq 1)$$

$$\frac{dp_0(t)}{dt} = \gamma(t)p_1(t). \tag{4}$$

Consumed with the usual terminology of probability theory, as in **Chapter 9** of Bailey (1990) [1], Eq. (4) is referred to as a *time inhomo-genous* or *non-homogenous birth-and-death* process, on account of the probability coefficients, $\alpha(t)$, $\beta(t)$ and $\gamma(t)$, being time-dependent quantities. These quantities "meter" the birth-and-death rates in Eq. (4), that is, the probability for a random process t experience both positive and negative jumps. Therefore, Eq. (4) is essentially a system of *differential-difference equations* in which the full dynamics of a linear birth-and-death process is being described more formally as a *discontinuous* Markov process.

No topic is more cardinal to the nature of a linear or pure birth processes, than the *Poisson process*. This process is so widely covered in various aspects of biology, physics and telephone engineering, to name a few, that the reader can simply refer to any well-known authored book on the subject, such as Feller (1968) [2], for further insight.

Specifically, the case of a *pure* or *simple linear-birth process*, in which $\gamma(t) = 0$ in Eq. (4), is referred to as the *Yule-Furry distribution* (Bailey (1990) [1]), a distribution which deservedly earmarks both Yule's (1924) [3] attempt to describe the evolution of a species within a particular genus as well as Furry's (1937) [4] venture to quantify the "passage" or *flux* of high-energy electrons through lead – vital, at that time, for understanding *cosmic-rays* or "showers" impinging on earth. In contrast to the Yule-Furry distribution, are the supporters of the *Arley-Feller distribution* or processes, who purport to the notion of adding more realism to the previous work on Markov processes by allowing for death processes to occur, i.e., $\gamma(t) \neq 0$ in Eq. (4).

In fact, it is the past work by Arley (1943) [5] on cosmic-ray phenomena, which incidentally maintains the same theme as Furry's (1937) [4] earlier work on the same subject, that encouraged particle-disintegration type models to be looked upon favourably in probability theory as either a linear death process, or as linear birth-and-death processes, generally.

One can certainly trace the influence of birth-and-death processes from the one end as being due to the latest uses of statistical models in particle physics colliders, a trend that is, perhaps, undeniably due to Williams' (1974) [6] original but critical account of random processes in nuclear reactors, to the mathematical surveying of earth'secological system, at the other end, in which the migration and extinction rates of various species – ranging from the smallest land invertebrates to the largest aquatic mammals on earth – is decisively pinned-down as an evolutionary birth-and-death process (Bailey (1990) [1]).

Gardiner's exposition (1996) [7] is also as good a place as any to make the connection between the birth-and-death differential-difference equations, as specified by Eq. (4), and the Fokker-Planck chemical equations found in the physical and chemical sciences, since such equations are categorically known in this respect as the *Kolmogorov forward-differential* or *backward-differential equations*.

What is unmistakably clear, so far, is that the enormous popularity of the Kolmogorov Master equations in the literature is primarily derived from the pioneering works of Yule-Furry and Arley-Feller. Indeed, it is these very pioneers that were instrumental for bringing back into probability theory the significance of the *geometric* or *binomial distribution*, also called the *negative-binomial distribution*, which today has become one of the most widely-used classes of probability distribut-ions in stochastic and statistical analysis.

At the other end of the spectrum, the name A. N. Kolmogorov, who is regarded as one of Russia's most influential and leading mathematician on the subject, is almost synonymous with the study of birth-and-death equations. Perhaps, in many ways, Yule's (1924) [3] early transactions, which attempts to develop a "mathematical theory of evolution", is the major culprit for the enormous popularity devoted to the Kolmogorov Master equations in the Russian literature, which may be the reason why so many scientists in Russia tend to expend almost all their efforts into the probability theory of birth-and-death equations, in the hope that a new theory of evolution will soon emerge (see also the translated mathematical exposition by Kolmogorov and Yushkevich's (1998) [10] for a comprehensible account of famous Russian mathematicians.

Currently, these types of birth-and-death differential equations are seen more readily in the mathematical literature; for instance, see Driver's (1977) [8] monograph, as a type of *delay-differential equation*, which the reader will find has reached popularity as a new branch in mathematical and operations research, namely, *renewal theory*.

There is definitely a new dogma in the literature with respect to the birth-and-death equations of Eq. (4) that is very much responsible for the main trend of attention and focus being drawn toward the use of Master equations in science – since it is "tied-up" with everything from the evolution of species, as well as the most minute organisms, to the disintegration of the most elementary of all particles – which makes it all the worthwhile for the present author to reference the birth-and-death differential-difference equations of Eq. (4) as the *Kolmogorov Master equations for birth-and-death.*

Of course, the question of existence and uniqueness of the solution of the Kolmogorov Master equations in Eq. (4) has always been considered *curiously uneasy.* In general, the probabilities $p_n(t)$ contained in Eq. (4) at time $t = 0$ will depend on the actual state of the system, in the sense that $p_n(0) = a_n$, for some sequence a_n. Here, the reader can refer to **Chapters 5-7** of Driver (1977) [8] and Bailey (1990) [1] for further insight into the questions related to existence and uniqueness of differential-difference equations. But make no mistake here, since the challenge in probability theory, generally, is how does one actually manifest the initial condition in order to find a unique solution for the Kolmogorov Master equations?

In this chapter, the unique solution for the Kolmogorov Master equations in Eq. (4) will be pursued as well as solved exactly, subject to an initial condition that is realistically represented by the Poisson process, namely,

$$p_n(0) = \delta_{n,0} \qquad \text{where} \qquad \delta_{n,0} = \begin{cases} 1 & n = 0, \\ 0 & n \neq 0, \end{cases} \tag{5}$$

which is also considered to be valid mathematically, since the Markov chain formed by the probabilities $p_n(t)$ in Eq. (4) start from $n = 0$.

In the next section, however, the reader will see the introduction of a parametric co-ordinate and, more importantly, how this type of parameterization can be utilized effectively to derive and establish an exact and general solution to the Kolmogorov Master equations of Eq. (4). At this point, one can solicit for a direct comparison of the exact solution as derived below by the present author with that of D. G. Kendall's (1948) [8] highly acclaimed "exact" result for the very same system of birth-and-death equations (defined in Eq. (4)) – a result which Kendall himself, perhaps, may raise a doubt as to its truth, since his solution does not satisfy the Poisson initial condition of Eq. (5). As the reader will soon encounter, there are a few surprises in store with the actual parameterization of the Kolmogorov Master equations. The first, is that the Kolmogorov equations are invariant under rotations of the parametric co-ordinate which itself has unexpectedly uncovered a hidden symmetry. Consequently, to one's amazement, it is this unexpected symmetry that will actually lead the way to *unlocking* a wider number of exact solutions than previously known to exist for the Kolmogorov Master equations, as will be presented toward the end of section §2 of the present chapter.

Lastly, in section §3 of this chapter, is the discussion of the unique solution of the Kolmogorov Master equations and its consequences.

2. Parameterization of the Master Equations

§. *The Differential-Difference Equations Simplified.* For the Kolmogorov Master equations or the birth-and-death differential-difference equations specified in Eq. (4), assume that the solution $p_n(t)$ can be represented as

$$p_n(t) = \frac{\delta^n(t)w_n(t)}{n}, \qquad n = 1, 2, 3, \ldots \qquad (6)$$

with $\delta(t)$ and $w_n(t)$ taken as arbitrary functions of the time t, since this is optimistically intended to elicit from Eq. (4) a system of solvable difference equations.

To see if this is true, one substitutes $p_n(t)$ from Eq. (6) into the first line of Eq. (4), to get

$$\frac{1}{n}\frac{dw_n(t)}{dt} + \left[\dot{\delta}(t)/\delta(t) - \beta(t)\right]w_n(t) \qquad (7)$$
$$= \dot{\alpha}(t)\,w_{n-1}(t)\big/\delta(t) + \gamma(t)\delta(t)w_{n+1}(t).$$

One has the freedom here to elect a choice for $\delta(t)$, such that

$$\frac{\alpha(t)}{\delta^2(t)\gamma(t)} = -\kappa, \qquad \Rightarrow \qquad \delta(t) \equiv \pm i\,\sigma(t) = \pm i\sqrt{\frac{\alpha(t)}{\kappa\,\gamma(t)}},$$

with i as the pure imaginary complex number, and κ an arbitrary constant whose value, shown later below, has a fixed designation. As usual, the time derivative of $\delta(t)$ is denoted by the dot symbol, i.e., $\dot{\delta}(t) = d\sigma(t)/dt$, which itself is not actually required for arriving at the exact solution to the Kolmogorov Master equations of Eq. (4). Nevertheless, it has an interesting form to appreciate,

$$\dot{\delta}(t) \equiv \frac{\pm\, i\, \sigma(t)}{2}\left[\frac{\dot{\alpha}(t)}{\alpha(t)} - \frac{\dot{\gamma}(t)}{\gamma(t)}\right],$$

since it indicates that $\dot{\delta}(t)$ will vanish when the birth-and-death parameters $\alpha(t)$ and $\gamma(t)$ are simply constants.

Introducing two more time-dependent quantities,

$$r(t) = \pm\frac{i}{\gamma(t)\sigma(t)} \quad \text{and} \quad q(t) = r(t)\left[\dot{\delta}(t)\big/\delta(t) - \beta(t)\right], \qquad (8)$$

enables Eq. (7) to transform into

$$r(t)\frac{dw_n(t)}{dt} + n\,q(t)\,w_n(t) = -n\big(\kappa\, w_{n-1}(t) - w_{n+1}(t)\big), \qquad (9)$$

whose form, as will become more apparent shortly, will be used to forge a connection with the difference equations of another well-known polynomial – the *Chebyshev polynomials*.

§. *The Difference Equations of the Chebyshev Polynomials.* From **Chapter 5.7** of the mathematical monograph by Magnus *et al* (1966) [11], the differential-difference equations for the Chebyshev polynomials of the *first kind*, denoted by $T_n(z)$, can be expressed, in terms of the variable $z = \cos\theta$, by

$$\left(1 - z^2\right)\frac{dT_n(z)}{dz} = n\left[T_{n-1}(z) - z\,T_n(z)\right],$$

which inaugurated with the standard recurrence formula for $T_n(z)$:

$$T_{n+1}(z) = 2z\,T_n(z) - T_{n-1}(z),$$

allows for the differential-difference equations of the Chebyshev poly-

30

nomials $T_n(z)$ to be re-modified as

$$4\left(1-z^2\right)\frac{dT_n(z)}{dz} + 2nz\,T_n(z) = n\left(3T_{n-1}(z) - T_{n+1}(z)\right). \qquad (10)$$

One can see almost straight-away the close resemblance between Eq. (10) and the Kolmogorov Master equations reformulated in Eq. (9), which automatically makes the Chebyshev polynomials of the *first kind,* $T_n(z)$, a lead candidate to solving the Kolmogorov Master equations.

In fact, further consultation with **Chapter 5** of Magnus *et al* (1966) [11], shows that there exists a second independent solution that satisfies the difference equations of Eq. (10). This second independent solution is called the Chebyshev polynomials of the *second kind,* denoted by $U_n(z)$.

To prove that $U_n(z)$ is a second independent solution, first define $S_n(z)$ by

$$S_n(z) = \sqrt{1-z^2}\,U_n(z),$$

and then switch from the z variable to the θ coordinate, using $z = \cos\theta$, so that a functional form of the differential-difference equations of Eq. (10) satisfied by $S_n(z)$, can be suitably expressed in terms of the θ variable as:

$$4\sin\theta\,\frac{dS_n(\cos\theta)}{d\theta} - 2n\cos\theta\,S_n(\cos\theta) = $$
$$= -n\left[3S_{n-1}(\cos\theta) - S_{n+1}(\cos\theta)\right]. \qquad (11)$$

It is worth noting here that the differential-difference equations drawn-out here for the Chebyshev polynomials, $S_n(z)$ and $T_n(z)$, in terms of the θ coordinate, are precisely the functional forms that one seesk after inorder to solve the Kolmogorov Master equations exactly. This will be confirmed next.

§. *General Solution of the Kolmogorov Master Equations.* All that needs to be done here, is to convert the modified form of the Kolmogorov Master equations of Eq. (9) into its equivalent functional form, mandated by Eq. (11). The following parameterization:

$$r(t)\dot{\theta}(t) = 4\sin\theta(t) \quad \text{and} \quad q(t) = -2\cos\theta(t), \tag{12}$$

will suffice, since, in conjunction with the usual chain rule $d/dt = \dot{\theta}(t)\,d/d\theta(t)$, this parameterization will automatically transpose the Kolmogorov Master Equations of Eq. (9) into

$$4\sin\theta(t)\frac{dw_n(t)}{d\theta(t)} - 2n\cos\theta(t)w_n(t) = -n\big(\kappa\, w_{n-1}(t) - w_{n+1}(t)\big). \tag{13}$$

Comparing this differential-difference equation with Eq. (11), and provided that one supplies the arbitrary constant on the right-hand side of Eq. (13) with the value $\kappa = 3$, it follows that the solution $w_n(t)$ corresponds exactly with the Chebyshev polynomials of the first and second kind, given by $T_n(z)$ and $S_n(z)$, respectively.

Physically, the parametric co-ordinate $\theta(t)$, adopted in Eq. (12), contains all the necessary information of a linear birth-and-death process, so that it is safe to say that the *dynamics* for all three of birth-and-death parameters, $\alpha(t)$, $\beta(t)$ and $\gamma(t)$, have truly been parameterized via the coordinate $\theta(t)$.

One can also determine $\theta(t)$ explicitly. First divide the parametric relations for $r(t)$ and $q(t)$, which are defined in Eq. (12), to obtain a first-order differential equation for $\theta(t)$:

$$\cot\theta(t)\dot{\theta}(t) = -2\big(\dot{\delta}(t)/\delta(t) - \beta(t)\big).$$

This relation has the basic first integral,

$$\ln\big(\sin\theta(t)\big) = -2\int\big(\dot{\delta}(\tau)/\delta(\tau) - \beta(\tau)\big)d\tau$$

$$= -2\ln\delta(t) + 2\int_0^t \beta(\tau)\,d\tau + c,$$

where c is a constant of the integration, with $\delta(t)$ defined in the relation given below Eq. (7).

Solving above for $\theta(t)$, explicitly gives

$$\theta(t) \equiv \arcsin z(t)$$
$$= -i \ln\left(iz(t) + \sqrt{1 - z^2(t)}\right),$$

(14)

where

$$z(t) = \frac{C e^{2\int_0^t \beta(\tau)\,d\tau}}{\delta^2(t)},$$

(15)

with the former integration constant c being re-absorbed by another constant, C. Eq. (15) merely reinstates the precise dependence of $\theta(t)$ with the birth-and-death parameters, since $z(t)$ can be seen to depend explicitly on them, which is also implicitly implied by Eq. (14).

There is no further doubt, the parameterization procedure advanced in the preceding paragraphs to solve for the Kolmogorov Master equations of Eq. (9), indeed, reveals that there are in fact two independent and exact solutions for $w_n(t)$.

So, without any further delay, one expresses the exact solutions $w_n(t)$ for the Kolmogorov Master equations of Eq. (9) as a linear combination of the Chebyshev polynomials of the first and second kinds, $T_n(\cos\theta(t))$ and $S_n(\cos\theta(t))$, resepectively,

$$w_n(t) = A S_n(\cos\theta(t)) + B T_n(\cos\theta(t)),$$

(16)

for arbitrary constants A and B.

From the result above for $w_n(t)$ in Eq. (16), together with the definition of $p_n(t)$ in Eq. (6), it follows that the most general and exact form of the solution for $p_n(t)$ is:

$$p_n(t) = \frac{(\pm i)^n \sigma^n(t)}{n}\left\{A S_n(\cos\theta(t)) + B T_n(\cos\theta(t))\right\}, \quad (n \geq 1),$$

(17a)

and

$$p_0(t) = K + \int_0^t \gamma(\tau) p_1(\tau)\, d\tau$$

$$= K \pm i \int_0^t \gamma(\tau) \sigma(\tau) \left\{ A \sin\theta(\tau) + B \cos\theta(\tau) \right\} d\tau, \tag{17b}$$

with K being another constant of the integration for $p_0(t)$, and where explicit expressions for the Chebyshev polynomials of degree $n = 1$, from pp. 256 of **Chapter 5.7** of Mangnus *et al* (1966) [11], have been inserted under the integral sign in the second-line of Eq. (17b) for the $p_1(t)$ probability.

§. *Reflective Properties of the Chebyshev Polynomials.* There are several useful properties of the Chebyshev polynomials contained in Eq. (17) worth exploiting here.

First, observe that under the transformation $\theta(t) \to \pi \pm \theta(t)$:

$$T_n\big(\cos\big(\pi \pm \theta(t)\big)\big) = (-1)^n\, T_n\big(\cos\theta(t)\big),$$

$$S_n\big(\cos\big(\pi \pm \theta(t)\big)\big) = (-1)^{n-1}\, S_n\big(\cos\theta(t)\big),$$

which indicates the sign reversal property associated with the Chebyshev polynomials, as can be verified from Table 22.4 on pp. 777 of Ambramowitz and Stegun (1964) [12], with the cosine property, $\cos\big(\pi \pm \theta(t)\big) = -\cos\theta(t)$.

Of course, the same transformation $\theta(t) \to \pi \pm \theta(t)$, gives $\sin\big(\pi \pm \theta(t)\big) = \mp \sin\theta(t)$, which conforms again to the sign reversal property above. This proves that the differential-difference equations in Eq. (13) satisfied by the $w_n(t)$'s in Eq. (16) are genuinely invariant under the transformation $\theta(t) \to \pi \pm \theta(t)$.

Since each of the independent solutions apprehend themselves in Eq. (17) as exact solutions of the Kolmogorov Master equations of Eq. (4), even under the transformation $\theta(t) \to \pi \pm \theta(t)$, it is found convenient to distinguish the transformed solutions $p_n\big(\pi \pm \theta(t)\big)$ by the labels:

$$p_n^{(s)}\left(\pi \pm \theta(t)\right) = \frac{\left(\pm i\right)^n \delta^n(t)}{n}\, A\, S_n\left(\cos\left(\pi \pm \theta(t)\right)\right),$$

$$p_n^{(T)}\left(\pi \pm \theta(t)\right) = \frac{\left(\pm i\right)^n \delta^n(t)}{n}\, B\, T_n\left(\cos\left(\pi \pm \theta(t)\right)\right).$$

$$(18)$$

There is a substantial difference between these transformed solutions and their original counterparts, $p_n^{(s)}\left(\theta(t)\right)$ and $p_n^{(T)}\left(\theta(t)\right)$, i.e., without invoking the transformation.

Though difficult to foresee, it is definitely true, nevertheless, that the arguments of the the Chebyshev polynomials will represent – without fail – quite different *orbits* and *trajectories* under the transformations $\theta(t) \to \pi \pm \theta(t)$. One verifies this fact from the definitions of the Chebyshev polynomials on pp. 256 of Mangus *et al* (1966) [11],

$$S_n\left(\cos\left(\pi \pm \theta(t)\right)\right) = \sin n\left(\pi \pm \theta(t)\right),$$
$$T_n\left(\cos\left(\pi \pm \theta(t)\right)\right) = \cos n\left(\pi \pm \theta(t)\right),$$

as, indeed, being totally separate and distinct continuous-time functions, respectively, to their counterparts, $S_n\left(\cos\theta(t)\right)$ and $T_n\left(\cos\theta(t)\right)$.

It turns out that this same argument, as with the transformation $\theta(t) \to \pi \pm \theta(t)$, equally applies also if one swings the transformation $\theta(t) \to \pi \pm \theta(t)$ around a full-circle (clockwise), like so $\theta(t) \to -\pi \pm \theta(t)$, so then $p_n^{(s)}\left(-\pi \pm \theta(t)\right)$ and $p_n^{(T)}\left(-\pi \pm \theta(t)\right)$ will naturally form totally separate classes of exact solutions for the Kolmogorov Master equations of Eq. (4), in addition to those already contained in Eq. (17).

Here one ploughs on-ahead to take full advantage of the rotational invariance property discussed in the previous paragraphs, by forming sums and differences between the transformed classes of solutions, $\theta(t) \to \pi \pm \theta(t)$ and $\theta(t) \to -\pi \pm \theta(t)$, and their counterparts (labeled by appropriate superscripts

$$^{\pm \pi - \theta} u_n(t) = p_n^{(s)}\left(\theta(t)\right) - p_n^{(s)}\left(\pm \pi - \theta(t)\right),$$
$$^{\pm \pi - \theta} v_n(t) = p_n^{(T)}\left(\theta(t)\right) - p_n^{(T)}\left(\pm \pi - \theta(t)\right).$$

$$(19)$$

This same conviction holds quite readily for the other set of linear combinations, namely,

$$
\begin{aligned}
{}^{\pm\pi+\theta}u_n(t) &= p_n^{(s)}\big(\theta(t)\big) + p_n^{(s)}\big(\pm\pi + \theta(t)\big), \\
{}^{\pm\pi+\theta}v_n(t) &= p_n^{(T)}\big(\theta(t)\big) - p_n^{(T)}\big(\pm\pi + \theta(t)\big),
\end{aligned}
\tag{20}
$$

which can again be confirmed to satisfy the Kolmogorov Master equations of Eq. (4).

It is worth re-iterating that Eqs. (19) and (20) are each in their own right unique solutions to the Kolmogorov Master equations of Eq. (4), and that, in general, such linear combinations will not cancel one another out under time evolution, since their *orbits* $\theta(t) \to -\pi \pm \theta(t)$, as argued earlier, generally follow different *trajectories* at arbitrary times t.

§. ***Anti-symmetrization and Uniqueness.*** It is not so absurd to realize that the linear combinations in Eqs. (19) and (20), which are formed from taking sums and differences between the respective transformed solutions, $p_n^{(s,T)}\big(\pm\pi + \theta(t)\big)$ and $p_n^{(s,T)}\big(\pm\pi - \theta(t)\big)$, and untransformed ones, $p_n^{(s,T)}\big(\theta(t)\big)$, is really no different to *anti-symmetrization of the wavefunction* in quantum mechanics; a technique that the present author is on most familiar territory with from his previous work on quantum scattering formalism for *exchange scattering* reactions.

A better acquaintance with anti-symmetrization and its renown *selection* rules under *parity* and *inversion* of the co-ordinates, as well as how it is used to establish uniqueness to the scattered wavefunction in quantum mechanics, can be gotten from the discussion on pp. 404 of **Chapter 11** of Davydov (1965) [13] (see also pp. 338 of **Chapter 10** of [13]).

In this respect, uniqueness of the solutions to the Kolmogorov Master equations is sustained in Eqs. (19) and (20) entirely by anti-symmetrization via their "mirror-like" and "inversion" properties with respect to rotations and reflections of the parametric co-ordinate $\theta(t)$, which can also readily be understood in terms of the schematics for "exchange scattering of orbits" in quantum mechanics employed on pp. 405 of **Chapter 11** of Davydov's (1965) [13].

It is equally important, though, to recognize that the linear combinations in Eqs. (19) and (20) were initially constructed with the view point of satisfying the Poisson initial condition of Eq. (5) at $t = 0$.

To prove that the initial condition of a Poisson process is satisfied by the anti-symmetrized solutions, Eqs. (19) and (20), one simply launches all orbits $\theta(t)$ at time $t = 0$ with the value

$$\theta(0) = \frac{\pi}{2}, \tag{21}$$

so then the integration constant $C\,(\neq 0)$, implicit in $\theta(t)$ through the quantity $z(t)$ in Eq. (15), becomes firmly fixed at $C = \delta^2(0)$. It then follows that all the linear combinations, as prescribed by Eqs. (19) and (20), namely,

$$^{\pm\pi-\theta}u_n(0) = {}^{\pm\pi-\theta}v_n(t) = {}^{\pm\pi+\theta}u_n(0) = {}^{\pm\pi+\theta}v_n(t) = \delta_{n,0}, \tag{22}$$

automatically satisfy the initial condition for the Poisson process, provided the second integration constant K for the probability $p_0(t)$ in Eq. (17) is assigned the value of $K = 1$.

One can now rest whole-heartedly with the results of this section – finally – since all admissible classes of solutions to the Kolmogorov Master equations of Eq. (4), subject to the Poisson initial condition $\delta_{n,0}$ being imposed, have now been fully uncovered.

3. Final Comments on the Exact Solutions & Outlook

Having come this far, it is worth reminding, that the parameterization method, elaborated at some lenght in section §2 of the present chapter, is based on a previous publication by the present author in Daniel (2004) [14], and so is rightly being hailed in this monograph as the first published work on the exact solution (i.e., true solution) of the Kolmogorov Master equations of Eq. (4) for birth-and-death processes.

By contrast more detail on the parameterization method has

has been devoted in section §2 of the present chapter than originally given in author's publication in *J. Phys. Soc. Jpn.* [14], with the essential difference that the present treatment focuses more toward under-pinning the uniqueness question attached to the Kolmogorov Master equations, a crucial point that was never addressed or proven by the present author, nor is it contained in the author's original solution set given by Eqs. (21) and (22) of Daniel (2004) [14].

A few typographical errors have been noticed in the published article by Daniel (2004) [14]. The first involves the solutions in Eqs. (21) and (22) of the article by Daniel (2004) [14], which is to be compared with those given in Eq. (17) of this chapter. Here one sees that there is a slight misprint which resulted in an omission of a factor in Eq. (22) of the published article [14], which has now been corrected under the integral sign in the third line of Eq. (17) of the present chapter, i.e., now includes the contributing factor $\pm i\sigma(t)$, where $\sigma(t)$ is identified in the text of this chapter below Eq. (7). The other glaringly obvious misprint is the factor of a $\frac{1}{2}$ included in the expression for $\delta(t)$ directly above Eq. (8) of the present chapter, but is absent in Eq. (6) of [14]. This is in not in any way a critical expression, and so does not alter the state of the exact solutions to the Kolmogorov Master equations obtained in the article by Daniel (2004) [14].

Before one overlooks the parameterization method adopted in the present chapter to solve the Kolmogorov Master equations for birth-and-death, via a parametric co-ordinate $\theta(t)$, it must be said that the real pride of the method lies in its ability to have *unclothe* a hidden symmetry inherent in the Kolmogorov Master equations. Indeed, the invariance property of the Kolmogorov Master equations with respect to rotations and reflections of the parametric $\theta(t)$ coordinate is very much considered a stunning shock; a result that was never foreseen at the start of this work.

Moreover, with so many influential contributors responsible for shaping and modernizing the Kolmogorov Master equations for birth-and-death processes, particularly in Russia, where the name A. N. Kolmogorov is already know to have such a deep impact in probability theory and statistical physics, it is not so incredulous then to expect the exact solutions to the Kolmogorov Master equations to be named in honor of another Russian mathematician (Chebyshev), as specified below respectively by the Chebyshev polynomials of the *first* and *second kind*, from Eq. (18),

$$p_n^{(s)}\big(\theta(t)\big) = \frac{(\pm i)^n\, \delta^n(t)}{n}\, A\, S_n\big(\cos\theta(t)\big),$$

$$p_n^{(T)}\big(\theta(t)\big) = \frac{(\pm i)^n\, \delta^n(t)}{n}\, B\, T_n\big(\cos\theta(t)\big),$$

There are two further points worth recollecting here: The first, is that the solutions above for the Kolmogorov Master equations represent the first evidence of a reciprocal dependence on the population index, n, which makes the above expressions additionally a rare-type of probability distribution to wonder. The second point, is that the present work generalizes the non-homogenous birth-and-death equat-ions in Eq. (4) to include the extra probability coefficient, $\beta(t)$, whereas in Kendall's (1948) [9] paper, the birth-and-death equations are always rigidly balanced and fixed in terms of two time-dependent birth-and-death parameters, i.e., $\beta(t) = -\big(\alpha(t) + \gamma(t)\big)$. This versatility, however, has yet to have been explored fully by the present author.

It is inevitable, though, that having now obtained a wider and more admissible classes of exact solutions to the Kolmogorov Master equations, presently available in Eqs. (19) and (20) of this chapter, that one can expect higher demands to be imposed toward correcting or curing completely many of the worldly problems – ranging, for instance, from nuclear physics to incurable diseases and epidemics, or even to earth's greater ecology of species. To put it succinctly, any previous misuse or deficiency with stochastic birth-and-death equations can now expect to be rectified with lesser risks

Finally, for peace of mind, the reliance on the Kolmogorov Master equations will always continue to prompt future generations of researchers to follow the one latitude of thought of advancing remedies and cures for the future, as it no doubt did with Yule (1924) [3] in his endeavour for a mathematical theory of evolution.

REFERENCES

1. N. T. J. Bailey, "The Elements of Stochastic Processes with Applications to the Natural Sciences," Wiley, New York (1990).

2. W. Feller, "An Introduction to Probability Theory and its Applications," John Wiley & Sons, New York (1968).

3. U. G. Yule, "A Mathematical Theory of Evolution based on the Conclusions of Dr. J. C. Willis", FRS, *Philosophical Transactions of the Royal Society of London*, **213**, 21-87 (1924).

4. W. H. Furry, "On Fluctation Phenomena in the Passage of High Energy Electrons through Lead", *Physics Review*, **52**, 569-581 (1937).

5. N. Arley, "On the Theory of Stochastic Processes and their Application to the Theory of Cosmic Radiation," Wiley, New York (1948).

6. M. M. R. Williams, "Random Processes in Nuclear Reactors," Pergamon Press, London (1974).

7. C. W. Gardiner, "Handbook of Stochastic Methods for Physics, Chemistry and the Natural Sciences," Springer-Verlag, New York (1996).

8. R. D. Driver, "Ordinary and Delay Differential Equations," Springer-Verlag, New York (1977).

9. D. G. Kendall, "On the Generalised 'Birth and Death' Process," *The Annals of Mathematical Statistics*, **19**, 1-15 (1948).

10. A. N. Kolmogorov and A. P. Yushkevich, "Mathematics of the 19th Century Function Theory according to Chebyshev, Ordinary Differential Equations, Calculus of Variations, Theory of Finite Differences," Birkhäuser Verlag, Basel, (1998).

11. W. Magnus, F. Oberhettinger and R. P. Soni, "Formulas and Theorems for the Special Functions of Mathematical Physics," (3rd Ed.), Springer, Berlin (1966).

12. M. Abramowitz and I. Stegun, "Handbook of Mathematical Functions : with Formulas, Graphs, and Mathematical Tables," Washington, U.S. Govt. Print. Off. (1964).

13. A. S. Davydov, "Quantum Mechanics," Pergamon Press, Oxford (1965).

14. D. J. Daniel, "On a Class of Birth-and-Death DDE's," *Journal of the Physical Society of Japan*, **73**, No. 7, 2028-2029 (2004).

15. E. A. McCulloch and J. E. Till, "Perspectives on the Properties of Stem Cells,"*Nature Medicine*, 11, No.10, 1026-1028 (2005).

CHAPTER 3

Galileo's Simple Pendulum Without Elliptic Functions

The *simple pendulum* is first reviewed from an historical point of view by the present author, and then, to one's astonishment, solved exactly, using only high-school algebra and known transcendental functions in the derivation.

1. Introduction

§. *A Brief History of the Simple Pendulum.* Galileo Galilei's (1564-1642) fascination with motion and dynamics of bodies, as recollected by Seeger in Galilei and Seeger (1966) [1], which is based on dialogues taken from Galileo's scientific testament, the *Two New Sciences*, stands out as one of his most outstanding achievements, because he had challenged the Aristotolean beliefs upon which the foundations of physics, mathematics, astronomy and philosophy had been built on at that time. Galileo's perceptions in science not only exceeded many of the Aristotolean ideas of that time but his inventions are known also to have definitely dawned a new technological era in Rome in the 17th century – such that this most remarkable Italian scientist can very well be coined the "father of technology". In fact, from **Chapter 2** of Galilei and Seeger (1966) [1], one envisages Rome during that time to have been the centre of science in Europe, and that Galileo's fascination with oscillations around 1590 would only marvel the world over, since his experiments and observations with "measuring time", as based on the notion of mechanical "swings" and "vibrations", would naturally enter our world as the everyday time-keeping devices that today's modern society refer to as clocks and watches.

Around this time can be traced the earliest records of Galileo's concept of a *simple pendulum* in which the motion of a suspended bob is seen to undergo *natural oscillations*, which one learns from **Chapter 15** of Galilei and Seeger (1966) [1] is based on Galileo's legendary observations of a swinging chandelier in the Cathedral of Pisa. It was in fact is these observations that led to Galileo's famous pendulum law, which Galileo himself formulated as the *law of isochronism*, meaning equality of time – because a pendulum will "sweep" back and forth at the same rate irrespective of the size or amplitude of its swing.

In its most primitive form, Galileo used pendulum bobs made of wooden objects or corks, as well as lead balls (e.g., see pp. 17 of Fermi and Bernardini (1961) [2]), for his simple pendulum experiments. Nevertheless, Galileo's simple pendulum can still be simulated perfectly, as shown in **Figure 1**, by using a light string (or weightless rod) of constant length $\mathbf{OA} = \ell$, which is suspended from one end at $\mathbf{O}$ and having a "bob" of mass m attached to the other end, but at the same time, is assumed to be freely "falling" under the influence of gravity g in a vertical plane.

Initially, at time t seconds, the bob is assumed to be moving with speed v, with its angle between the string and the downward vertical $\mathbf{OA}$ at an angle θ radians. To frame this precisely, one has at $t = t_0$ seconds, $v(t_0) = v_0$ and $\theta(t_0) = \theta_0$ radians. The conservation of energy equation for the simple pendulum, accordingly, is then given by the relation

$$\tfrac{1}{2}mv_0^2 + mg\ell\cos\theta_0 = \tfrac{1}{2}mv^2 + mg\ell\cos\theta,$$

or

$$\tfrac{1}{2}\left(v_0^2 - v^2\right) = g\ell\left(\cos\theta_0 - \cos\theta\right). \tag{1}$$

Resolving tangentially along the arc $\mathbf{AP} = s = \ell\theta$, one has the defining relations $\omega_0^2 = g/\ell$ and $v(t) \equiv ds/dt = \ell\, d\theta(t)/dt$, which can be incorporated into Eq. (1) as

$$\left(\frac{d\theta(t)}{dt}\right)^2 = 2\omega_0^2\left[\frac{v_0^2}{2g\ell} - \cos\theta_0 + \cos\theta(t)\right]. \tag{2}$$

One can easily verify this equation as the first integral for the *pendulum's equation of motion* :

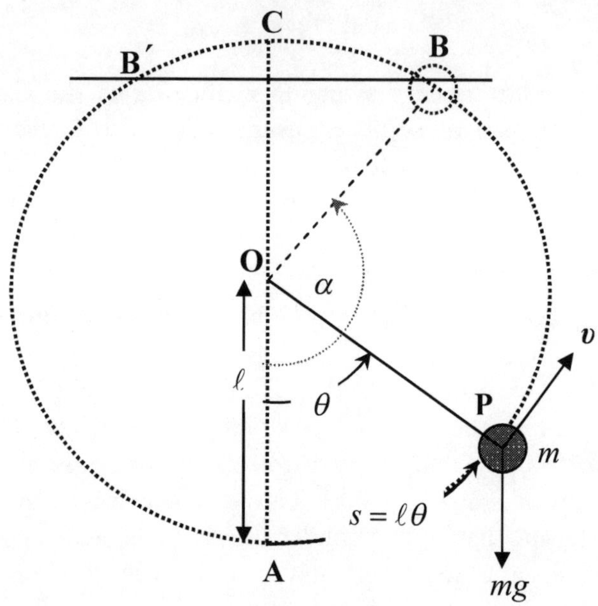

Figure 1. The Simple Pendulum

$$\frac{d^2\theta(t)}{dt^2} + \omega_0^2 \sin\theta(t) = 0 \, , \tag{3}$$

with ω_0 being the angular frequency of oscillation.

Now, if one glances back at Eq. (2) and takes the square-root on the right-hand side, one notices

$$\frac{d\theta(t)}{\sqrt{\dfrac{v_0^2}{2g\ell} - \cos\theta_0 + \cos\theta(t)}} = \pm\sqrt{2}\,\omega_0\,dt \, , \tag{4}$$

which means Eq. (2) is separable in $\theta(t)$ and t. It is furiously impossible, as all past efforts indicate, to integrate Eq. (4) directly, since it cannot be solved with any of the elementary or transcendental functions known in mathematics.

Quite rightly, then, a more sophisticated approach in the mathematics literature had to be developed for the accurate determination of the pendulum's equation of motion, as briefly reviewed next.

§. ***Mathematical Theory of Elliptic Functions.*** It is customary to assume that $\theta(t_0) = \theta_0 = 0$ radians and also use the trigonometric identity $2\sin^2\frac{1}{2}\theta = 1 - \cos\theta$ under the radical sign appearing in the denominator of the left-hand side of Eq. (4). Also, let

$$k^2 = \frac{v_0^2}{4g\ell}, \tag{5}$$

then Eq. (4) can be rewritten as

$$\frac{d\theta(t)}{\sqrt{k^2 - \sin^2\frac{1}{2}\theta(t)}} = \pm\sqrt{2}\,\omega_0\,dt.$$

The physical basis for k^2 in Eq. (5) can be made clearer, if one makes a change of the variable from $\theta(t)$ to $\varphi(t)$ as follows,

$$\sin\tfrac{1}{2}\theta(t) = k\sin\varphi(t) \equiv \sin\tfrac{1}{2}\alpha\sin\varphi(t), \tag{6}$$

as this will establish that $k^2 = \sin^2\frac{1}{2}\alpha$. In terms of the new variable $\varphi(t)$, Eq. (4) now reads,

$$\frac{d\varphi(t)}{\sqrt{1 - k^2\sin^2\frac{1}{2}\varphi(t)}} = \pm\omega_0\,dt. \tag{7}$$

Geometrically, the angle α is the angular amplitude of oscillations for the simple pendulum (referring to **Figure 1**), since it represents the maximum angle through which the pendulum will elevate along the arc **AB**. The full angle of oscillation for $\theta(t)$ is, of course, 2α, indicated in the figure by **B′OA + AOB**. As a consequence $\theta(t)$ oscillates between $\pm\alpha$, which means the range of oscillation of $\varphi(t)$ in Eq. (6) lies between $\pm\frac{1}{2}\pi$. In this instance, $d\theta(t)/dt$ in Eq. (4) changes sign according to the direction of the swing of the pendulum. In particular, if the pendulum swings from the left-to-right, i.e., counter-clockwise from $\mathbf{B}'$ to $\mathbf{B}$, taken here as the positive direction, then $d\theta(t)/dt$ must be of opposite or negative sign when the pendulum swings from the right-to-left, i.e., clockwise from $\mathbf{B}$ to $\mathbf{B}'$. This sign change in $d\theta(t)/dt$ is logically expected on physical grounds, though the present author has found this to be totally ignored in the relevant literature, which asserts instead that $d\theta(t)/dt$ changes sign according to whether $\theta(t)$ is $\pm\alpha$. There is also the tendency for certain authors to use the notation $\theta_0 = \alpha$, such as in pp. 69 of Barger and Olsson (1995) [3], for representing the maximum angle of oscillation, simply because the maximum angle, denoted by $\theta_{\max} = \alpha$, physically corresponds to the situation where the final speed of the pendulum has reached its terminal velocity of $v(t) = 0$ in the energy relation of Eq. (1). When this happens, the terminal velocity is determined as $v_0^2/2 = g\ell(1 - \cos\theta_{\max}) \equiv 2g\ell\sin^2\frac{1}{2}\alpha$, so that the reason for asserting the relation $k^2 = \sin^2\frac{1}{2}\alpha \equiv v_0^2/4g\ell$ earlier above should now be more at ease with the reader.

It is easy to convince the reader here that Eq. (4) (or Eq. (7)) cannot be reduced further by any of the elementary, algebraic, circular or hyperbolic functions, so according to the mathematics literature one assumes instead that there exists a function $F(\varphi, k)$ which solves Eq. (7) as

$$\omega_0 t = \int_{\varphi(0)}^{\varphi(t)} \frac{d\varphi(t)}{\sqrt{1 - k^2\sin^2\frac{1}{2}\varphi(t)}} \equiv F(\varphi, k). \tag{8}$$

This integral equation is formally, if not historically, the "rendezvous" point for a mathematical theory of elliptic functions. This has been impressed upon, for example, in **Chapter 1** (as well as the Preface) of Greenhill (1959) [4], which the treatment below will follow to some extent.

§. *The Jacobian Elliptic Functions.* The integral defined by $F(\varphi, k)$ in Eq. (8) is called an *elliptic integral* of the *first kind*, with *amplitude* $\varphi(t)$ and *modulus* k. Eq. (8) is an elliptic integral of the first kind in *standard form*, since it was first employed by Adrien-Marie Legendre in his treatise, *Theorie des fonctions elliptiques*, published in 1825.

Actually, the theory of elliptic functions is recognized as having begun, more or less, with the treatises by Niels Hendrik Abel, *Œuvres*, published in 1841, and Carl Gustav Jacobi, *nova theoriae functionum ellipticarum*, published in 1829, which seals the theory of elliptic functions to a period saw Abel and Jacobi as the two lead mathematicians in the field.

In practice, elliptic integrals of the kind defined by Eq. (8) are recognized whenever integrals containing cubic or higher-order polynomials $R(x)$ appear under a square-root sign, i.e., $\int \sqrt{R(x)}$. It is of course the presence of the radical sign associated with elliptic integrals that makes them amenable to being solved formally as an extension of the algebra associated with trigonometric functions, as will be shown next.

Abel's remark in 1823 to stricken the meaning of $F(\varphi,k)$ as an inverse function, by putting $u = F(\varphi,k)$, is considered quite revolutionary, since one can look into the properties of the amplitude, $\varphi(t)$, as a function of u. That is, $\varphi = F^{-1}(u,k)$, rather than u as a function of $\varphi(t)$, as originally adopted by Legendre in his treatise, Theorie des fonctions elliptiques. It would be dishonest here to ignore the popularity gained by Jacobi's notation, $\varphi = \operatorname{am} u$, or $\varphi = \operatorname{am}(u,k)$, for the amplitude, since this notation brought a new form of elegance into the theory of elliptic functions.

In this sense, Jacobi could see three elliptic functions associated with u,

$$u = F(\varphi,k) = \int_{\varphi(0)}^{\varphi(t)} \frac{d\varphi(t)}{\sqrt{1 - k^2 \sin^2 \frac{1}{2}\varphi(t)}},$$

with modulus satisfying $0 < k^2 < 1$, namely:

$$\cos \varphi(t) = \cos\big(\operatorname{am}(u,k)\big), \quad \sin \varphi(t) = \sin\big(\operatorname{am}(u,k)\big),$$
$$\text{and } \Delta\varphi(t) = \Delta\big(\operatorname{am}(u,k)\big).$$

A more abbreviated notation of these elliptic functions, which is in present use, however, came slighlty later and follows the suggestion used by Christoph Gudermann in his treatise, *Theorie der Modular*

Functionen, published in 1839; *viz:*

$$\mathrm{cn}\, u = \cos \varphi(t) = \cos\big(\mathrm{am}(u, k)\big),$$
$$\mathrm{sn}\, u = \sin \varphi(t) = \sin\big(\mathrm{am}(u, k)\big), \qquad (9)$$
$$\mathrm{dn}\, u = \Delta\varphi(t) = \Delta\big(\mathrm{am}(u, k)\big);$$

referred to as the *Jacobian elliptic functions.*

Starting out, that is, algebraically, from the derivative properties associated with the Jacobian elliptic functions:

$$\frac{d\,\mathrm{cn}\, u}{du} = -\mathrm{sn}\, u \, \mathrm{dn}\, u, \qquad \frac{d\,\mathrm{sn}\, u}{du} = \mathrm{cn}\, u \, \mathrm{dn}\, u, \qquad \frac{d\,\mathrm{dn}\, u}{du} = -k^2 \mathrm{sn}\, u \, \mathrm{cn}\, u, \quad (10)$$

and

$$\frac{d\varphi(t)}{du} = \sqrt{1 - k^2 \sin^2 \tfrac{1}{2}\varphi(t)} = \mathrm{dn}\, u \,, \qquad (11)$$

together with the definitions of the three elliptic functions given by Eq. (9), it is easily shown that the elliptic integral in Eq. (8) will be sufficiently satisfied by the Jacobian elliptic functions of Eq. (9).

This shows that the problem of solving the equations of motion of the simple pendulum as having been delegated to the mathematical problem of *inversion* of the integral in Eq. (8). For the present purposes, inversion of the integral for the simple pendulum will be carried out explicitly below, which follows the treatment described, for example, in **Chapter 1** of Greenhill (1959) [4] and also **Chapter 3** of Bowman (1953) [5], for three separate cases of the simple pendulum's equation of motion, beginning first with the definition of the period of the pendulum.

§. *The Pendulum Period and Solutions.* The period of the pendulum refers to the time of a *double swing*, and coincides with the pendulum bob elevating from **A** to **B**, and then sweeping back from **B** to **B′**, before descending again from **B′** to the start at **A**, where it assumed that the bob in **Figure 1** started out its motion at **A**. If one denotes the pendulum's period here by T, then, it is clear that the time of motion of the pendulum from **A** to **B** will be only *one-quarter* of the period T, or $\tfrac{1}{4}T$. Furthermore, since the initial increase in time t, from t_0 to $\tfrac{1}{4}T$ seconds, corresponds to an increase in $\theta(t)$ from

θ_0 to α radians, the excursion of $\varphi(t)$ in Eq. (6), as a consequence, spans only from 0 to $\frac{1}{2}\pi$. Accordingly, $\omega_0(t-t_0)$ or u in Eq. (8) ranges from 0 to K, and so, by coincidence, leads to a new integral quantity,

$$K = \int_{\theta_0}^{\frac{1}{2}\pi} \frac{d\varphi(t)}{\sqrt{1-k^2\sin^2\frac{1}{2}\varphi(t)}}; \tag{12}$$

called Legendre's *complete elliptic integral* of the *first kind*. Physically, the complete elliptic integral defined here is a true measure of the real *quarter period* of the pendulum.

Case A: Oscillatory Motion $\left(v_0^2/4g\ell < 1\right)$. Initially, $\theta_0 = 0$ and one's attention is confined to the motion of the pendulum swinging with finite amplitude from one side to the other side about **A** in **Figure 1**. This is a case of pure oscillatory motion, which constrains the modulus defined in Eq. (5) to $k^2 = v_0^2/4g\ell < 1$. By further adopting the relationship between $\theta(t)$ and $\varphi(t)$ in Eq. (6) as well as the definitions for the three Jacobian elliptic functions, as enumerated in Eq. (9), one can write

$$\sin\tfrac{1}{2}\theta(t) = k\sin\varphi(t) = k\,\mathrm{sn}(\omega_0 t, k),$$

$$\cos\tfrac{1}{2}\theta(t) = \sqrt{1-k^2\sin\varphi(t)} = \mathrm{dn}(\omega_0 t, k),$$

where the dependence on the modulus k is made explicit here in the arguments of the elliptic functions. It is now a matter of applying the derivative properties of the Jacobian elliptic functions prescribed in Eqs. (10) and (11), together with the above algebraic identities. The problem of a simple pendulum undergoing finite oscillations is then explicitly solved as

$$\theta(t) = 2\sin^{-1}\left(k\,\mathrm{sn}\,\omega_0 t\right), \quad \text{and} \quad \frac{d\theta(t)}{dt} = 2k\omega_0\,\mathrm{cn}\,\omega_0 t,$$

where the modulus k has been suppressed in the arguments of the elliptic functions. The pendulum's complete period T is naturally determined by $\omega_0 T = 4K$, or $T = 4K\sqrt{\ell/g}$.

Case B: Circulatory Motion $\left(v_0^2/4g\ell > 1\right)$. In this case, the pendulum bob is assumed to impart from **A** $(\theta_0 = 0)$ with sufficient velocity that enable it to perform complete revolutions about the vertex **C** in **Figure 1**. In this case, the derivative $d\theta(t)/dt$ will always possess the same sign, and, moreover, since the modulus is defined as $k^2 = 4g\ell/v_0^2$, it can be shown that

$$\left(2\omega_0/k\right)dt = \frac{d\theta(t)}{\sqrt{1 - k^2\sin^2\frac{1}{2}\theta(t)}},$$

after a slight rearrangement of $d\theta(t)/dt$ in Eq. (11). Again, resorting to the definition and derivative properties of the elliptic functions, given respectively by Eqs. (6) and (11), one finds

$$\sin\tfrac{1}{2}\theta(t) = \sin\varphi(t) = \operatorname{sn}\left(\frac{\omega_0 t}{k}\right), \quad \text{and} \quad \cos\tfrac{1}{2}\theta(t) = \cos\varphi(t) = \operatorname{cn}\left(\frac{\omega_0 t}{k}\right),$$

so then the equations of motion of a simple pendulum undergoing circulatory motion is now captured as

$$\theta(t) = 2\sin^{-1}\left(\operatorname{sn}\left[\frac{\omega_0 t}{k}\right]\right) \quad \text{and} \quad \frac{d\theta(t)}{dt} = \frac{2\omega_0}{k}\operatorname{dn}\left(\frac{\omega_0 t}{k}\right),$$

with the complete period T of the pendulum determined by $\omega_0 T/k = 2K$, or $T = 2kK\sqrt{\ell/g}$.

Case C: Vertical Position $\left(v_0^2/4g\ell = 1\right)$. This case corresponds to a situation when the pendulum bob imparts from **A** at $\theta_0 = 0$, but only has just enough velocity to reach the vertex **C** of the circle in **Figure 1**. In this instance, one clearly has $k^2 = 4g\ell/v_0^2 \equiv 1$, which reduces the integral of Eq. (8) into a well-known standard integral, with the result

$$\sin\tfrac{1}{2}\theta(t) = \tanh\omega_0 t \quad \text{and} \quad \cos\tfrac{1}{2}\theta(t) = \operatorname{sech}\omega_0 t,$$

Here the situation of the pendulum's bob never actually reaching the vertical position in **Figure 1,** can be seen almost straight-away, since the equations of motion of the pendulum in this case are completely governed by hyperbolic functions,

$$\theta(t) = 2\sin^{-1}\left(\tanh \omega_0 t\right) \quad \text{and} \quad \frac{d\theta(t)}{dt} = 2\omega_0 \operatorname{sech} \omega_0 t .$$

This now completes the mathematical preliminaries associated with elliptic integrals and their related properties.

Historically, the framework of *pendulum motion* described in the present section stems from the conceptual notion of *parameterization of the ellipse*. This is because elliptic functions were originally defined in the context of a *rectification* around the perimeter of an ellipse. The modulus k then, naturally endows the meaning of the *eccentricity* of the ellipse, to which **Chapter 3** of Bowman (1959) [5] should be referred to for the details.

One more important point worth mentioning, is the *double-periodicity* associated with the Jacobian elliptic functions of Eq. (9), which is the result of the pendulum's period being an imaginary number. According to the insight of Appell, see **Chapter 4** of Whittaker (1937) [6], the *double-periodicity* associated with the Jacobian elliptic functions can also be proven to emerge as a consequence of **reversing the action of gravity** in the pendulum problem, which defines the pendulum's period of motion as a pure imaginary quantity.

2. Exact Solution of the Pendulum's Equation of Motion

In this section, the exact solution for the simple pendulum is produced, without recourse to any kind of sophisticated mathematical procedures or the need of special functions in mathematics. So, the remarkable feat of solving the equations of motion of the simple pendulum exactly, accomplished in this monograph before your very eyes, does not at all extend beyond the capabilities of some simple trigonometric identities, accompanied with some elementary algebra.

§. *Formulation of the Exact Solution of the Simple Pendulum.*
For the simple pendulum, the initial conditions, namely, its initial angle

θ_0 (in radians) and initial velocity v_0, are given by

$$\theta_0(t_0) = \theta_0 \quad \text{and} \quad \left.\frac{d\theta(t)}{dt}\right|_{t=t_0} = v_0, \tag{13}$$

respectively. The derivative $d\theta(t)/dt$ in Eq. (4) is first rewritten more compactly as

$$\frac{d\theta(t)}{\sqrt{\cos\theta(t) - \lambda_0}} = \pm\sqrt{2}\,\omega_0\,dt, \tag{14}$$

with $\lambda_0 = \cos\theta_0 - v_0^2/2g\ell$ assigned.

A rather splendid step, is to next assume that the solution $\theta(t)$ of the simple pendulum has an exact form of the kind specified by

$$\theta(t) = \sec^{-1}\left(\frac{\delta x(t) + a}{x(t) + b}\right), \tag{15}$$

in which $x(t)$ is an auxiliary function of the time t to be determined, and δ, a and b are arbitrary constants.

Next one recalls the following identity for the inverse secant function,

$$\frac{d\sec^{-1}x}{dx} = \frac{1}{|x|\sqrt{x^2 - 1}} \equiv \frac{1}{x^2\sqrt{1 - \dfrac{1}{x^2}}},$$

taken from Eq. 4.4.56, pp. 82 of Abramowitz and Stegun (1964) [7]. Subsequent use of this identity for $\theta(t)$ in Eq. (15) now allows the derivative on on the left-hand side of Eq. (14) to be written as ($\delta b \neq a$),

$$\frac{(\delta b - a)\,dx(t)}{\sqrt{(\delta x(t) + a)\left[(1 - \lambda_0\delta)x(t) + (b - \lambda_0 a)\right]}} \tag{16}$$

$$= \pm\sqrt{2}\,\omega_0\,\sqrt{\left[(\delta - 1)x(t) + a - b\right]\left[(\delta + 1)x(t) + a + b\right]}\,dt$$

It is an elementary matter to recognize that the substitution for $\theta(t)$ in Eq. (15) has allowed the derivative $d\theta(t)/dt$ on the left-hand side of Eq. (14) to be cast into the form of a product of two quadratic polynomials, under a common square-root, which has been made more obvious in Eq. (16), with the result

$$\frac{(\delta b - a)\, dx(t)}{\varepsilon\sqrt{x^2(t) + \alpha x(t) + \beta}\,\sqrt{x^2(t) + \gamma x(t) + \eta}} = \pm\sqrt{2}\,\omega_0\, dt, \quad (\delta b \neq a) \quad (17)$$

where α, β, γ, η and ε are constants, whose precise forms will be revealed later below, since they can be shown to be are related to the coefficients, λ_0, δ, a and b, appearing in Eq. (15).

The reader will find it extremely rewarding to note that the integration on the left-hand side of Eq. (17) can be performed rather easily, simply by imposing the requirement $\alpha = \gamma$ and $\beta = \eta$. This is permissible since α, β, γ, η are arbitrary constants. Eq. (17) then transforms as

$$\frac{(\delta b - a)\, dx(t)}{\varepsilon\sqrt{x^2(t) + \alpha x(t) + \beta}\,\sqrt{x^2(t) + \gamma x(t) + \eta}} \equiv \frac{dx(t)}{\varepsilon\left(x^2(t) + \alpha x(t) + \beta\right)},$$

which leads to a well-known standard integral,

$$\therefore \int\frac{(\delta b - a)\, dx(t)}{\varepsilon\left(x^2(t) + \alpha x(t) + \beta\right)} = \frac{2(\delta b - a)}{\varepsilon\sqrt{4\beta - \alpha^2}}\,\tan^{-1}\left(\frac{2x(t) + \alpha}{\sqrt{4\beta - \alpha^2}}\right), \quad (18)$$

as can be verified with Eq. 3.3.16, pp. 12 of Abramowitz and Stegun (1964) [7]. The result in Eq. (18) is so elementary; in fact, the reader no longer needs to be ponder anxiously here over its validity.

There is, yet, another astonishing fact. Since the order of the terms linear in $x(t)$ appearing under the square root sign in Eq. (16) can be commuted and interchanged with one another under the radical signs, this means that three combinations of exact solutions for the simple pendulum exist in reality!

For purposes of keeping track of the combinations that arise when the terms linear in $x(t)$ are interchanged and commuted under the radical signs in Eq. (16), each particular combination will be enumerated by the letter subscript r, so that the constants α, β, γ, and η, in general, will carry r as a subscript.

With this enumeration in mind, the general result for the integral of Eq. (18) reads

$$\pm\sqrt{2}\,\omega_0(t-t_0)+c \equiv \pm\sqrt{2}\,\omega_0\int_{t_0}^{t}dt$$

$$= \frac{2(\delta b-a)}{\varepsilon\sqrt{4\beta_r-\alpha_r^2}}\,\tan^{-1}\!\left(\frac{2x(t)+\alpha_r}{\sqrt{4\beta_r-\alpha_r^2}}\right),\quad (r=1,2,3).\quad (19)$$

where c is an arbitrary constant of the integration.

The result in Eq. (19) means that the whole problem of solving for the pendulum's equation of motion is now within one's grasp. This can be done directly by a straight-forward inversion of the trigonometric function in Eq. (19), since this, in turn, determines $x(t)$ completely, which can then be substituted into Eq. (15) for $\theta(t)$.

The constant of integration, c, appearing on the left-hand side of Eq. (19) can be determined uniquely. Let $x(t)$ be prescribed with an initial value at time $t=t_0$, say $x(t_0)=x_0$, then, from the initial conditions of Eq. (13), one obtains

$$x_0 = \frac{a-b\sec\theta_0}{\sec\theta_0-\delta}\quad\text{and}\quad c=\tan^{-1}\!\left(\frac{2x_0+\alpha_r}{\sqrt{4\beta_r-\alpha_r^2}}\right).\qquad (20)$$

Substituting these expressions back into Eq. (19) and inverting, leads immediately to an explicit expression for $x(t)$,

$$x(t) = -\tfrac{1}{2}\alpha_r + \tfrac{1}{2}\sqrt{4\beta_r-\alpha_r^2}\times$$

$$\times\ \tan\!\left[\frac{\pm\varepsilon\sqrt{2}\,\omega_0\sqrt{4\beta_r-\alpha_r^2}}{2(\delta b-a)}(t-t_0)+\tan^{-1}\!\left(\frac{2x_0+\alpha_r}{\sqrt{4\beta_r-\alpha_r^2}}\right)\right],\quad (21)$$

where $\delta b \neq a$ is assumed, and $\varepsilon = \sqrt{(\delta^2 - 1)(1 - \lambda_0 \delta)}\,\delta$ holds through-out, for each $r = 1, 2, 3$, on account that this is a common term factored-out from under the radical signs in the denominator of the left-hand side of Eq. (16).

It then follows, from Eqs. (14) and (15), that the solution $\theta(t)$ and its derivative $d\theta(t)/dt$ can now be determined completely, respectively as

$$\theta(t) = \sec^{-1}\left(\frac{\delta x(t) + a}{x(t) + b}\right)$$

$$\frac{d\theta(t)}{dt} = \pm\sqrt{2}\,\omega_0 \sqrt{\frac{(1 - \lambda_0 \delta)x(t) + (b - \lambda_0 a)}{\delta x(t) + a}} \tag{22}$$

where $x(t)$ is explicitly given by Eq. (21).

One now arrives at the fact that Eq. (22) represents the exact solution for the equations of motion of the simple pendulum.

§. ***Determination of*** α, β, γ, *and* η. At this stage, all that remains is to carry out some algebra in order to relate the α, β, γ, and η variables more precisely to the original coefficients λ_0, δ, a and b appearing in Eq. (14). This will be delineated below in three parts, where $\delta b \neq a$ is assumed to hold throughout, unless otherwise stated.

Case 1: The first case starts by regrouping Eq. (16) into the form

$$\frac{(\delta b - a)}{\sqrt{(\delta x(t) + a)[(1 - \lambda_0 \delta)x(t) + (b - \lambda_0 a)]}} \times$$

$$\times \frac{dx(t)}{\sqrt{[(\delta - 1)x(t) + a - b][(\delta + 1)x(t) + a + b]}}$$

$$= \frac{(\delta b - a)\,dx(t)}{\varepsilon\sqrt{x^2(t) + \alpha_1 x(t) + \beta_1}\,\sqrt{x^2(t) + \gamma_1 x(t) + \eta_1}},$$

where, in order for this relation to hold, one requires that

$$\alpha_1 = \frac{a(1 - 2\lambda_0\delta) + b\delta}{(1 - \lambda_0\delta)\delta}, \quad \beta_1 = \frac{a(b - \lambda_0 a)}{(1 - \lambda_0\delta)\delta},$$

$$\gamma_1 = \frac{2\delta a - 2b}{\delta^2 - 1}, \quad \eta_1 = \frac{a^2 - b^2}{\delta^2 - 1}. \tag{23}$$

But from the preceding discussion, in connection with arriving at the integration result of Eq. (18), one has to also impose $\alpha_1 = \gamma_1$ and $\beta_1 = \eta_1$ in Eq. (23). This further requirement leads to:

$$\alpha_1 = \gamma_1 \quad \Rightarrow \quad (a - b\delta)\left(\delta^2 - 2\lambda_0\delta + 1\right) = 0,$$

$$\beta_1 = \eta_1 \quad \Rightarrow \quad a^2(\delta - \lambda_0) - b^2\delta(1 - \lambda_0\delta) - ab\left(\delta^2 - 1\right) = 0,$$

which, together with the assumption $\delta b \neq a$, produces two well-defined roots for each of the constants, δ and the ratio a/b,

$$\delta \equiv \delta_{1,2} = \lambda_0 \pm \sqrt{\lambda_0^2 - 1},$$

$$\frac{a}{b} = \frac{\left(\delta^2 - 1\right) \pm \sqrt{\left(\delta^2 - 1\right)^2 - 4\delta(1 - \lambda_0\delta)(\delta - \lambda_0)}}{2(\delta - \lambda_0)}. \tag{24}$$

Case 2: The next regrouping of terms linear in $x(t)$ under the square root sign of Eq. (16), produces

$$\frac{(\delta b - a)}{\sqrt{[(\delta - 1)x(t) + a - b][(1 - \lambda_0\delta)x(t) + (b - \lambda_0 a)]}} \times$$

$$\times \frac{dx(t)}{\sqrt{(\delta x(t) + a)[(\delta + 1)x(t) + a + b]}}$$

$$= \frac{(\delta b - a)\,dx(t)}{\varepsilon\sqrt{x^2(t) + \alpha_2 x(t) + \beta_2}\,\sqrt{x^2(t) + \gamma_2 x(t) + \eta_2}},$$

where now

$$\alpha_2 = \frac{(\delta - 1)(b - \lambda_0 a) + (a - b)(1 - \lambda_0\delta)}{(1 - \lambda_0\delta)(\delta - 1)}, \quad \beta_2 = \frac{(a - b)(b - \lambda_0 a)}{(1 - \lambda_0\delta)(\delta - 1)},$$

$$\gamma_2 = \frac{\delta(a+b) - a(\delta+1)}{\delta(\delta+1)}, \quad \eta_2 = \frac{a(a+b)}{\delta(\delta+1)}. \tag{25}$$

Imposing $\alpha_2 = \gamma_2$ and $\beta_2 = \eta_2$, again, leads to the set of algebraic equations:

$$\alpha_2 = \gamma_2 \;\Rightarrow\; (a - b\delta)\big[(1 + 2\lambda_0)\delta^2 + 1\big]\big[(1 + 2\lambda_0)\delta^2 - 2\delta - 1\big] = 0,$$

$$\beta_2 = \eta_2 \;\Rightarrow\; a^2\big(2\lambda_0\delta^2 - \delta + 1\big) - b^2\delta(\delta+1) + ab\big[(1 + 2\lambda_0)\delta^2 + 1\big] = 0,$$

whose roots, similarly, (as in **Case 1**) are

$$\delta \equiv \delta_{1,2} = \frac{1 \pm \sqrt{2(1 + \lambda_0)}}{(1 + 2\lambda_0)}, \tag{26}$$

$$\frac{a}{b} = \frac{-\big[(1 + 2\lambda_0)\delta^2 + 1\big] \pm \sqrt{\big[(1 + 2\lambda_0)\delta^2 + 1\big]^2 + 4\delta(\delta+1)\big(2\lambda_0\delta^2 - \delta + 1\big)}}{2\big(2\lambda_0\delta^2 - \delta + 1\big)}.$$

Case 3: For the final regrouping of the terms linear in $x(t)$ under the radical sign in Eq. (16), one gets

$$\frac{(\delta b - a)}{\sqrt{\big[(\delta+1)x(t) + a + b\big]\big[(1 - \lambda_0\delta)x(t) + (b - \lambda_0 a)\big]}} \times$$

$$\times \; \frac{dx(t)}{\sqrt{(\delta x(t) + a)\big[(\delta - 1)x(t) + a - b\big]}}$$

$$= \frac{(\delta b - a)\,dx(t)}{\varepsilon\sqrt{x^2(t) + \alpha_3 x(t) + \beta_3}\,\sqrt{x^2(t) + \gamma_3 x(t) + \eta_3}},$$

where

$$\alpha_3 = \frac{(\delta+1)(b - \lambda_0 a) + (a + b)(1 - \lambda_0\delta)}{(1 - \lambda_0\delta)(\delta+1)},$$

$$\beta_3 = \frac{a(1 - \lambda_0 - 2\lambda_0\delta) + b\big[2 + \delta(1 - \lambda_0)\big]}{(1 - \lambda_0\delta)(\delta+1)},$$

$$\gamma_3 = \frac{\delta(a-b)+a(\delta-1)}{\delta(\delta-1)}, \quad \eta_3 = \frac{a(2\delta-1)-b\delta}{\delta(\delta-1)}. \qquad (27)$$

In this case, invoking, $\alpha_3 = \gamma_3$ and $\beta_3 = \eta_3$, leads to the algebraic relations,

$$\alpha_3 = \gamma_3 \quad \Rightarrow \quad (a-b\delta)\left[(1-2\lambda_0)\delta^2 + 2\delta - 1\right] = 0,$$

$$\beta_3 = \eta_3 \quad \Rightarrow \quad a^2\left[(2\lambda_0-1)\delta\right] - b^2\delta(\delta-1) + ab\left[\delta^2+1\right] = 0,$$

with corresponding roots

$$\delta \equiv \delta_{1,2} = \frac{-1\pm\sqrt{2(1-\lambda_0)}}{(1-2\lambda_0)},$$

$$\frac{a}{b} = \frac{-\left(\delta^2+1\right)\pm\sqrt{\left(\delta^2+1\right)^2 - 4\delta(\delta-1)\left[(2\lambda_0-1)\delta-1\right]}}{2\left[(2\lambda_0-1)\delta-1\right]}. \qquad (28)$$

§. ***Practical Considerations.*** Since there will be two roots for δ, labeled above as $\delta_{1,2}$, and two corresponding roots associated with the ratio a/b, which is apparent from Eqs. (24)-(28), it is clear that a total of four separate solutions will result for each of the three cases considered above. In reality, then, one actually has **12 independent exact solutions!** for representing the pendulum's equation of motion.

From the present results, one sees that the solution described explicitly in Eq. (22), in general, encompasses the entire complex plane. This is because the nature of the roots associated with δ and a/b can become complex, depending on the pendulum's initial conditions. This is in direct contrast with the solution for the simple pendulum that emerged out with the theory of elliptic functions outlined in the Introduction, where there the solution based on elliptic function only admits real values. In practice, one also has the freedom of choosing a value for the constant b. In fact, this constant can be set to unity, $b = 1$ say, or, one can normalize b to match any of the denominators that appear in the roots $\delta_{1,2}$, as enumerated, respectively, by Eqs. (24), (26) and (28), for the three cases considered.

It is very elegant, of course, to be able to represent the various trajectories possible with the phase-space portrait of the simple pendulum, which one notices can be accomplished in the Jacobian elliptic functions through the modulus k, defined in Eq. (5), or equivalently, by using its angle of oscillation α (defined in Eq. (6)). In other words, the different values that α can assume shows that there exists a definite correspondence, i.e., parameterization, with the various energy surfaces of the simple pendulum.

To ensure that the same parameterization is possible in the present work, at least for the case of oscillatory motion, one simply makes the connection with the parameter λ_0 (see the line below Eq. (14)) and the angle of oscillation α, by using the relation

$$\lambda_0 = 1 - 2\sin^2 \tfrac{1}{2}\alpha, \tag{29}$$

where $v_0^2 / 4g\ell < 1$ is assumed to hold and $\theta_0 = 0$ radians at time $t = t_0$.

A further practical consideration that is hardly ever discussed for the simple pendulum, is that due to the expected sign changes which occur whenever the pendulum bob sweeps from left-to-right (counter-clockwise) or right-to-left (clockwise), so that what is needed here is a way of automatically correcting for the directional changes experienced by the pendulum. Fortunately, this can be done by "tailoring" a suitable factor that changes sign according to the direction of motion of the pendulum, which one anchors to both the solution $\theta(t)$ and its derivative $d\theta(t)/dt$ in Eq. (22). To be more specific, consider:

$$\theta(t) = (-1)^{T_{1/4}} \sec^{-1}\left(\frac{\delta x(t) + a}{x(t) + b}\right),$$

$$\frac{d\theta(t)}{dt} = (-1)^{T_{1/2}} \sqrt{2}\omega_0 \sqrt{\frac{(1 - \lambda_0\delta)x(t) + (b - \lambda_0 a)}{\delta x(t) + a}}, \tag{30}$$

where $T_{1/4}$ and $T_{1/2}$ correspond with the quarter and half periods of the pendulum.

These factors can be explicitly formulated as follows

$$T_{1/4} = \left[\frac{(t - t_0)}{K} \right] \quad \text{and} \quad T_{1/2} = \left[\frac{(t - t_0) + K}{2K} \right], \quad (31)$$

where [ ] is the greatest integer symbol, $K(2K)$ is the real quarter (half) period of the pendulum, and K the complete elliptic integral of the first kind, defined in Eq. (12).

Likewise, the auxiliary $x(t)$ part of the solution, determined previously by Eq. (21), has to be re-adjusted to read as

$$x(t) = -\frac{\alpha_r}{2} + \frac{\sqrt{4\beta_r - \alpha_r^2}}{2} \times$$

$$\times \tan\left[\frac{(-1)^p \, \varepsilon \sqrt{2}\,\omega_0 \sqrt{4\beta_r - \alpha_r^2}}{2(\delta b - a)} (t - t_0) + \tan^{-1}\left(\frac{2x_0 + \alpha_r}{\sqrt{4\beta_r - \alpha_r^2}} \right) \right], \quad (32)$$

where $p = T_{1/4} + T_{1/2}$.

The important point to recollect here is that the observed changes in direction of motion of the pendulum's bob, indicated by the $\pm$ signs, which one sees explicitly in governing equation for the pendulum's velocity $d\theta(t)/dt$, defined in Eq. (4), has to be accommodated for in any true solution for the pendulum. This makes it perfectly clear as to the reason for having introducing the period factors $T_{1/4}$ and $T_{1/2}$ in Eq. (31), since they are specifically tailored to accommodate for the directional changes associated with the velocity of the pendulum's bob – a matter that has never been touched upon in the literature.

3. Conclusion

It has been over four-centuries since the simple pendulum, as originally modeled on Galileo's legendary observations of a swinging chandelier, ever reached so glorious a moment as the present. This is self-explanatory, on account that the equations of motion of the simple pendulum have now, for the first time, been solved both exactly and analytically in the present monograph, as seen in the accompanying equations directly above. This rare feat is a total contrast with the development of Jacobian elliptic functions that were discussed in the introduction of this chapter, which until now were believed to be the only known solutions (approximate that is) possible for the equations of motion of a simple pendulum.

It is gladly said, that an important consequence of this work, is that cubic or elliptic integrals of the type

$$\int \frac{dx}{\sqrt{Ax^3 + Bx^2 + Cx + D}},$$

can now be evaluated without resorting to the Jacobian elliptic functions.

To prove this, first observe that the integrand in the above elliptic integral can be transformed by a suitable change of variables. That is, change from $x \to \varrho x' + \sigma$, where ϱ, σ are constants, and then change again to $x' = e^{\theta}$, which transforms the elliptic integral into:

$$\int \frac{dx}{\sqrt{Ax^3 + Bx^2 + Cx + D}} = \frac{1}{\sqrt{A\varrho}} \int \frac{dx}{x'\sqrt{x' + \dfrac{1}{x'} + \dfrac{(3\sigma A + B)}{A\varrho}}}$$

$$= \frac{-i}{\sqrt{2A\varrho}} \int \frac{d(i\theta)}{\sqrt{\cos i\theta - \lambda_0}}, \qquad \left(x' = e^{\theta}\right).$$

Closer inspection of the integrand above, reveals that σ has been chosen to satisfy the cubic equation $A\sigma^3 + B\sigma^2 + C\sigma + D = 0$, while the constant ϱ is chosen to satisfy $\varrho = \pm\sqrt{3\sigma^2 A + 2\sigma B + C/A}$, with $\lambda_0 = -(3\sigma A + B)/2A\varrho$.

One is now in a position to make the association between the integral shown on the second line above with the standard integral result of Eq. (19), which necessitates the desired result

$$\int \frac{dx}{\sqrt{Ax^3 + Bx^2 + Cx + D}} = \frac{-2i}{\varepsilon\sqrt{2A\varrho}} \frac{(\delta b - a)}{\sqrt{4\beta_r - \alpha_r^2}} \times$$

$$\times \tan^{-1}\left(\frac{e_r x^2 - 2f_r x + g_r}{\left[-\delta x^2 + 2(\delta\sigma + \varrho)x - \delta\left(\sigma^2 - \varrho^2\right)\right]\sqrt{4\beta_r - \alpha_r^2}} \right),$$

where the expressions for δ, ε, α_r and β_r correspond to the three cases already considered before in the previous section, as enumerated by $r = 1, 2, 3$. By following the algebra in that section, one can carefully show that the constants appearing in the argument of the inverse tangent function above satisfy the relations

$$e_r = 2a - \delta\alpha_r, \quad f_r = 2a\sigma + 4b\varrho - \delta\sigma\alpha_r - \varrho\alpha_r,$$

$$g_r = 2a\sigma^2 + 4b\varrho\sigma - 2a\varrho^2 - \delta\sigma^2\alpha_r - \delta\varrho^2\alpha_r.$$

To prop up the prospect of evaluating the elliptic integrals analytically, is a matter that no scientist, at first sight, would even conceive or "dream" possible for elliptic integrals – though the reader, having now reached this far, should feel the opposite or at least more inclined to accept this matter as a scientific truth.

REFERENCES

1. Galileo Galilei and R. J. Seeger, "Men of physics : Galileo Galilei, His Life and His Works," Pergamon Press, Oxford (1966).

2. L. Fermi and G. Bernardini, "Galileo and the Scientific Revolution," Basic Books, New York (1961).

3. V. D. Barger and M. G. Olsson, "Classical Mechanics: A Modern Perspecitve," McGraw-Hill, New York (1995).

4. G. Greenhill, "The Applications of Elliptic Functions," Dover Publications, New York (1959).

5. F. Bowman, "Introduction to Elliptic Functions with Applications," English Universities Press, London (1953).

6. E. T. Whittaker, "A Treatise on the Analytical Dynamics of Particles and Rigid Bodies, with an Introduction to the Problem of Three Bodies," Cambridge University Press, London (1937).

7. Abramowitz and I. Stegun, "Handbook of Mathematical Functions: with Formulas Graphs, and Mathematical Tables," Washington, U.S. Govt. Print. Off. (1964).

EPILOGUE

This monograph culminates three extraordinary problems in theoretical physics, all of which were virtually "handpicked" by the author of this monograph, based on the reputation that such problems possessed no known exact solutions, or even if exact solutions did exist for them, they were considered unattainable or at least impossible to obtain analytically.

The author's first choice to present the linear harmonic oscillator in Chapter 1 of this monograph is purposeful. Erwin Schrödinger's wavefunction solution for the simple harmonic oscillator is tacitly used in Chapter 1 to bring down a "myth" in quantum mechanics. The myth revolves around the fact that the linear harmonic oscillator is claimed to have been exactly and completely solved in quantum mechanics by Erwin Schrödinger, when if fact it hasn't. The logic to follow here is that the Schrödinger wave equation in quantum mechanics is founded on the principles that when the potential energy of the oscillator vanishes, the state of the wave function must reduce to the plane-wave solutions that would ordinarily be met with in free-space. Unfortunately, Erwin Schrödinger's solution for the simple harmonic oscillator fails to become a plane-wave when the potential energy of the oscillator vanishes, which is made explicit in Chapter 1 by allowing the limit of the angular frequency of the oscillator to approach zero, i.e., $\lim \omega \to 0$.

This very same failure is seen also in the hydrogen atom as well as for many other related problems in quantum mechanics. Consequently, since many of the known problems in quantum mechanics, fail to incorporate the plane-wave solutions in the total description of the wave function, one is left with very little option but to conclude that practically all wave function calculations in quantum mechanics, to date, can hardly be correct.

Again, the objective of the first chapter is to produce the exact wave function for the harmonic oscillator, but bear in mind that there was a price to pay before arriving at the exact result displayed in Chapter 1 of this monograph, since a three-term recursion relation had to be confronted as well as solved exactly. And unless the reader the reader feels challenge or motivated enough to close the infinite sums in

Eq. (21) of that chapter, the exact wave function will persist as an infinite product of Weber's function and the Laguerre function.

In the second problem, the author has appealed to the cunningness used in the early developments of quantum mechanics, for purposes of solving the Kolmogorov Master equations for birth-and-death. In this instance, the principles of anti-symmetrization and exchange scattering are invoked in order to establish uniqueness amongst all the possible exact solutions obtained in Chapter 2 of this monograph for the Kolmogorov Master equations for birth-and-death, so that there are now ten independent exact solutions for the Kolmogorov Master equations.

With ten solutions for the Kolmogorov Master equations now practically left waiting at the "altar" of Chapter 2 of this monograph, there is no telling of what one can expect to be unearthed in science the near future. In particular, whether there is a problem out there where all ten solutions will be required, can only remain a mystery at this stage.

The last and most illustrious problem, which is used to set the stage for the third problem in Chapter 3 of this monograph, is Galileo's simple pendulum. To state the obvious, the simple pendulum has always been regarded as one of the most difficult of all known nonlinear dynamical systems in nature to accurately describe, and just about everyone, from the greatest scientific minds and mathematicians of the sixteenth and seventeenth centuries to present, have all had to pay their respects to Jacobi's mathematical theory of elliptic functions as being the closest to an exact solution for the simple pendulum.

Having mentioned so many "giants" in Chapter 3 of this monograph for their part in the overall development of the theory of elliptic functions, it was, therefore, quite a bold move for the author to have been able to draw out an exact solution of the simple pendulum in Chapter 3 of this monograph without recourse to Jacobi's elliptic functions. In fact, it has by now almost become a tearful reunion, since practically all the foundations and sophistication in mathematics and physics, that had so mightily been laid out for describing the equation of motion of the simple pendulum, have now virtually been torn down and relegated to the level of high school algebra, as readers can witness for themselves in Chapter 3.

Despite the pendulum as an age-old problem, the exact solution of the simple pendulum obtained in this monograph should never be perceived as being outdated, since it is undeniably known in nonlinear science that there are so many more mysteries, than meets the eye, that one can probe for in nature with pendulum type equations. Of course, whether scientists are now one step closer to nature with the pendulum solution produced in this monograph, is probably the most anxious question on everyone's lips, so to speak.

Made in the USA
Charleston, SC
06 November 2010